Ecocriticism and Indigenous Studies

This book addresses the intersections between the interdisciplinary realms of ecocriticism and Indigenous and Native American studies and between academic theory and pragmatic eco-activism conducted by multi-ethnic and Indigenous communities. It illuminates the multilayered, polyvocal ways in which artistic expressions render ecological connections, drawing on scholars who work in collaboration with Indigenous artists from all walks of life and in all fields—including film, literature, performance, and other forms of multimedia—to expand existing conversations. Both local and global in its focus, the volume includes essays from multi-ethnic and Indigenous communities across the world, visiting topics such as Navajo opera, the histories of Sámi film production, South Indian tribal storied practices, Māori art installations, Native American and First Nations science fiction literature and film, Amazonian poetry, and many others. Highlighting trans-Indigenous sensibilities that speak to worldwide crises of environmental politics and action against marginalization, the collection alerts readers to movements of community resilience and resistance, cosmological thinking about inter- and intra-generational multispecies relations, and understandings of Indigenous aesthetics and material ecologies. It engages with emerging environmental concepts such as multispecies ethnography, cosmopolitics, and trans-Indigeneity, as well as with new areas of ecocritical research such as material ecocriticism, biosemiotics, and media studies. In its breadth and scope, this book promises new directions for ecocritical thought and environmental humanities practice, providing thought-provoking insight into what it means to be human in a locally situated, globally networked, and cosmologically complex world.

Salma Monani is Associate Professor in the Environmental Studies Department at Gettysburg College, USA.

Joni Adamson is Professor of Environmental Humanities, Department of English, and Senior Sustainability Scholar, Global Institute of Sustainability, Arizona State University, USA.

Routledge Interdisciplinary Perspectives on Literature

For a full list of titles in this series, please visit www.routledge.com.

60 **Tropical Gothic in Literature and Culture**
The Americas
Edited by Justin D Edwards and Sandra Guardini Vasconcelos

61 **The Contemporaneity of Modernism**
Literature, Media, Culture
Edited by Michael D'Arcy and Mathias Nilges

62 **The Comic Imagination in Modern African Literature and Cinema**
A Poetics of Laughter
Maik Nwosu

63 **Transnational Narratives from the Caribbean**
Diasporic Literature and the Human Experience
Elvira Pulitano

64 **Cartographies of Exile**
A New Spatial Literacy
Edited by Karen Elizabeth Bishop

65 **The Contemporary Literature-Music Relationship**
Intermedia, Voice, Technology, Cross-Cultural Exchange
Hazel Smith

66 **Post-Conflict Literature**
Human Rights, Peace, Justice
Edited by Chris Andrews and Matt McGuire

67 **Landscape, Seascape, and the Eco-Spatial Imagination**
Edited by Simon C. Estok, I-Chun Wang, and Jonathan White

68 **Auto/Biography across the Americas**
Transnational Themes in Life Writing
Edited by Ricia Anne Chansky

69 **Ecocriticism and Indigenous Studies**
Conversations from Earth to Cosmos
Edited by Salma Monani and Joni Adamson

Ecocriticism and Indigenous Studies

Conversations from Earth to Cosmos

Edited by Salma Monani
and Joni Adamson

Routledge
Taylor & Francis Group
LONDON AND NEW YORK

First published 2017 by Routledge

2 Park Square, Milton Park, Abingdon, Oxfordshire OX14 4RN
52 Vanderbilt Avenue, New York, NY 10017

Routledge is an imprint of the Taylor & Francis Group, an informa business

First issued in paperback 2020

Library of Congress Cataloging-in-Publication Data

Names: Adamson, Joni, 1958– editor. | Monani, Salma editor.
Title: Ecocriticism and indigenous studies: conversations from earth to cosmos / edited by Joni Adamson and Salma Monani.
Description: New York; London: Routledge, 2016. | Series: Routledge interdisciplinary perspectives on literature; 69 | Includes bibliographical references and index.
Identifiers: LCCN 2016013296
Subjects: LCSH: Environmental protection and motion pictures. | Indigenous peoples and mass media. | Ecology in motion pictures. | Indians of North America in motion pictures. | Documentary films—History and criticism. | Performing arts—Environmental aspects. | Indigenous peoples—Intellectual life. | Indigenous peoples—Ecology. | Ecocriticism.
Classification: LCC PN1995.9.E78 E36 2016 | DDC 809/.9336—dc23
LC record available at https://lccn.loc.gov/2016013296

ISBN: 978-1-138-90297-8 (hbk)
ISBN: 978-0-367-66812-9 (pbk)

Typeset in Sabon
by codeMantra

Salma: To those who told me my first stories and to hoopoe birds, elephant gods, Babugoosha the ant, and other beings who peopled such story worlds.

Joni: To the brilliant Barbara A. Babcock, who initiated me into the world of the liminal, the betwixt and between, and revealed to me the reasons why bears are good to think.

Contents

List of Figures

Foreword

Simon J. Ortiz

Each chapter of *Ecocriticism and Indigenous Studies: Conversations from Earth to Cosmos* asks the question, "Is modern industrial progress destructive to Indigenous communities and cultures?" This is a question I've asked many times in a lifetime of experiences I've had as an Indigenous American. And also because it's a question that relates to "Men on the Moon," a short fiction story I wrote more than 25 years ago that I'll mention in this Foreword (Ortiz 1999, 3–14).

As an *Aacquumeh* boy, I grew up on the Acoma Indian Reservation through which a little river—called *chunah* in our native Keres language—flowed eastward for a distance of forty miles from the Zuni Mountains and the Continental Divide to the west. Entering Acoma lands at Anzac, the *chunah* flowed past *Aacqumeh* farming and pasture lands in *Deetseyaamah*, a village where the *Aacqumeh hanoh*—Acoma people—had gardens, fruit orchards, and hay fields and lived in stone and adobe-walled homes scattered west, east, and north of the village.

The little *chunah*, regarded as a living entity or earth-being by many people, was a briskly flowing and sparkling clear stream of water gushing through and pouring alongside jagged formations of black lava rock. The *chunah* streamed through channels and cleavages the water carved in the rock for endless geologic time, 9,000 years or more. It flowed eastward eighty-five more miles towards the Manzano Mountain foothills where it spilled into the Rio Grande that flowed southward toward the Gulf of Mexico hundreds of miles away.

The *chunah,* flowed along the northern boundary of our homelands designated by the U.S. government as "Indian reservation," and was one of the prettiest little rivers in the Southwest. Well, it was before the massive uranium mining and refining industry began in the early 1950s in the region that became known as the Grants Uranium District.

By the time I was a teenager, the river had changed drastically as a result of the social-economic-political developments in the region that sometime seem totally incomprehensible. And, *haa-uh*—yes—at Anzac near the western border of the Acoma reservation, one could see the *chunah* had turned into a grayish liquid that seemed to have slowed down to a languid, sluggish, and syrupy grayish flow due to the uranium mining and refining industry upstream.

***

When I think about the *chunah*, my short story, "Men on the Moon," featured in the "Introduction" to *Ecocriticism and Indigenous Studies*, comes to mind. The story involves Faustin, elder grandpa to Amarosho, his grandson. Faustin tries to guide his grandson to be aware environmentally. He advises him to take care of the earth, our mother. She gave us our lives and provides for our care. This is a traditional Indigenous belief.

One day, Faustin and his grandson are watching professional wrestling on TV. The show is soon over when "Apache Red" wins. Then a shot of the U.S. space mission to the moon comes on next. Amarosho tells Faustin that American men called astronauts are going to the moon in a space ship. He points to the space rocket on the TV screen. The old man stares at a sharp-nosed cylinder emitting smoke from its lower end. Faustin thinks his grandson is joking with him at first. But Amarosho has been to Indian School where he learned strange things. His grandson is serious. "Look, Grandpa, a rocket will take men to the moon!" Faustin is skeptical for good reason. The night before, old Faustin had dreamt about Flintwing Boy, an important being in *Aacquumeh* culture. Later, he tells Amarosho about the dream because it holds important knowledge. In the dream, a *Sqkuyuu mahkina*, or giant monster machine, is destroying all the plants, animals, and waters the land contains. Coyote—*Tsuushki*—comes running to Flintwing Boy. He is very frightened, and he cries, "What is that?"

Flintwing Boy says, "It's some kind of *Sqkuyuu*." Calmly, he takes sacred corn pollen from his medicine prayer bag and puts some in Tsuushki's hand. They pray to earth spirits all around them, thanking them for spiritual knowledge, protection, guidance. They offer cornmeal to the land all around. As he fits an arrow to his bow, Flintwing tells *Tsuushki* to go tell the *hanoh*, or people, he is going to stop the *Sqkuyuu* that is destroying life. "Go and tell the people. I am going forward to meet it."

***

Believing. Seeing. Knowing. Awareness. For me, growing up in a relatively thriving Indigenous Acoma culture was a condition and circumstance unmatched. We lived the Indigenous life around us. We believed in the life we were born into. We lived life as learned from our parents and elders who were our teachers, protectors, human guides, and who taught us an Indigenous Acoma way of life that was a social-cultural guidance system.

In the morning, when you wake and rise, thank the Oshrahtrah as it is rising in the east. Thank the Dawn that is happening. Sunlight at Dawn and throughout each day helps us all. Be thankful, always appreciate light provided you and your sisters and brothers and all your relatives.

Our prayers were literally spoken as oral tradition instruction given to us by our parents and grandparents. This was guiding knowledge by which "to see" and "to know."

We believed in the way of life within the *Aacqumeh* world. We saw with eyes formed, trained, impacted, and influenced by the life of our extended families around us. As an infant, I had no other social, cultural, and physical context in which I learned the realities of my life.

A kind of "be all and end all" association is realized by what an infant sees and experiences. "Knowing" is initiated in a dynamic having to do with the association made between what is seen and what is felt by touching. An idea? A thought? Speculation? Osmosis? Infants respond to cuddling, gentle massaging, caressing by a parent. One can see movement within infants' eyes and faces having to do with "knowing." In "Men on the Moon," this is the *knowing* Faustin wants his grandson to be aware of always.

Awareness comes from knowing. That's what we see moving in the eyes and facial comportment of infants as they snuggle in the arms of their mothers. Security, comfort, warmth, food, gentleness is assured. *Confirmation, reassurance, affirmation* and *knowing* are associated with my story about Flintwing Boy, Coyote, and cornmeal. They come to mind when I think about the *chunah* that flowed gently and quietly near our *Deetseyaamah* home. The *chunah* was reassurance. I had no doubt the *chunah* as an earth-being would always be there. For sustenance. For sustainability. I *knew* it would be there. There was no reason to expect it would not be there. Just like the infant child has every reason to believe and expect his or her mother's breast will be there to provide nourishment.

In the *Aacqumeh* cultural world, even the way of praying is a reflection of that belief and practice: *we thank you, beloved sacred holy beings, for the safety we have; our health and protection are in your keep; we seek protection within your guiding knowledge; your sacred knowledge is our security. Knowing* then is a necessary and intimate responsiveness to the powers of the earth and the environmental terrain within and upon which we reside.

In Indigenous tradition, secular spirituality is ordained and certain. Awareness takes place as we grow. We grow and we change. We're aware of our growth. We grow in knowledge. The *chunah*, an earth-being, sacred and a necessity, is a natural, earthly phenomenon and an organic fact of life. As a boy, I went to the *chunah* countless times; it was unquestionably part of the *Aacqumeh* existence and my guide as I grew towards maturity. At the same time, I continued to gain awareness. Knowledge was the door that was being opened for me.

Two days after high school graduation, in the early 1960s, I applied for a job at the geological labs at Kerr McGee (Kermac), a petroleum corporation with oil exploration and drilling in Oklahoma, and later Texas, and the Gulf of Mexico. Almost immediately, I was hired as a chemical lab assistant. I worked as a lab assistant only for a week or so before I found out Kermac workers in the refinery's crushing, leaching, and drying sections were promoted faster than in other jobs at Kermac. So despite ambitions and plans to be a chemistry major in college, I opted to earn more money by working to become a machinery or equipment operator in crushing, leaching, or drying.

By then, the *chunah* was negatively affected by the tailings and waste effluents of the refinery process that was now spread over a large region by several large refineries. These substances included powerful acids and other chemicals that were very destructive, detrimental, and unhealthy to plant, animal, and human life. When they enter and pollute the water drainage systems, the natural environment is affected.

When the first major uranium mine called Jackpile Mine opened and began to operate in the early 1950s on Laguna Pueblo land near the village of Paguate, the *Kawaikahmehteetra*—Laguna people—had no idea of the dangers involved. But persuasive speakers, Bureau of Indian Affairs (BIA) officials, mining resource personnel, and others, had promised there would be employment opportunities galore; high wages would be earned; jobs as mechanics, heavy equipment operators, truck drivers, electricians, high explosive technicians, engineers, book keepers and other office jobs would be available to the local population. Job training and acquisition of employable skills were part of the persuasive spiel Laguna Pueblo tribal officials were given in order to gain their assent. Simply, it was the BIA spokespeople and mining corporation front men in cahoots with each other that talked the Laguna tribal leaders into accepting the plans to open Jackpile Mine. People needed jobs, and that's the argument that won the day: Jobs meant work for pay.

***

What my thinking comes down to has to do with the critical issue of modern industrial progress and Indigenous traditional existence. I've narrated a range of my own experiences and also reflected on the experiences of Indigenous communities with which I'm most familiar, Acoma and Laguna, neighboring sister tribes living next door to each other. Many other Indigenous tribal peoples in the U.S. and throughout the Americas are knowledgeable about the experiences of other communities and cultures in the world being targeted for economic and social development. Many have recounted the devastating and traumatic changes and experiences spoken of above.

In *Ecocriticism and Indigenous Studies: Conversations from Earth to Cosmos*, readers will find the devastation and trauma I have described above occurring time and again in the places, communities, and among the peoples discussed in the chapters of the book. I cannot help but think of Flintwing Boy and Coyote watching the *Sqkuyuu mahkina* destroying plant, animal, and all natural life. There is no choice but to do what Flintwing Boy decides to do: step forward to stop the destruction of the sacred land, the beautiful living entity upon which life depends. This is also what Faustian wants his beloved grandson Amarosho to know. This is the only way we, as a responsible human society, can save our precious sparkling rivers, Planet Earth, and ourselves!

Simon J. Ortiz
March 1, 2016
Tempe, Arizona

Ortiz, Simon. 1999. "Men on the Moon." *Men on the Moon: Collected Short Stories*. Tucson: University of Arizona Press. 3–14.

Acknowledgments

This book owes its presence to many. We owe much to the Indigenous communities whose visions grace the pages of this book. We are grateful for and honored by the generosity with which they have shared their work, their thoughts, and their wisdom with us. This collection would not have been possible without them, or the many others who have inspired us through our years working on amazingly creative collaborations between Indigenous peoples, artists, writers, environmental humanists, scientists, filmmakers, and our own teachers.

We are deeply indebted to the insights of our many colleagues—both at our local institutions (Gettysburg College and Arizona State University) and through our planetary networks (e.g., the Rachel Carson Center for Environment and Society and the Humanities for the Environment Observatories). We owe much to the editorial team at Routledge. A special thanks to editors Elizabeth Levine, Assunta Petrone, and Katherine Gilbert for their patience and guidance.

We could not do this work without the amazing support of our immediate families. Nor could we do it without the passions of our students. They inspire us to present this book as a means of remembering hidden pasts, understanding contemporary challenges, and envisioning alternative futures of eco-social hope.

Introduction

Cosmovisions, Ecocriticism, and Indigenous Studies

Joni Adamson and Salma Monani

In 2015, the Earth experienced its hottest year on record ("State of the Climate," 2016), and nations from around the world sent representatives to the United Nations' 2015 Conference on Climate Change in Paris, France, to negotiate an agreement to address climate change. The warming climate has caused many international politicians, artists, film stars, academics, activists and ordinary citizens alike, to question what the future holds. Many scientists are referring to these changes with a neologism, "the Anthropocene," first proposed in a short essay by Paul Crutzen, a Nobel-prize winning atmospheric chemist, and Eugene Stoermer, an ecologist, to designate a new post-Holocene epoch marked by human-caused changes so vast that they are being considered a geomorphic force of planetary scale (Crutzen and Stoermer 2000).[1] Both climate change and growing eco-social injustices aggravated by it have sparked a number of popular press stories about the future of Earth, and quite a few movies and television series speculating about possibilities for colonizing and terraforming other planets when Earth's systems can no longer support life. *The Martian* (2015), starring Matt Damon as a human attempting to grow food in the hostile, dry soils of the red planet, is only the latest Hollywood blockbuster to speculate about whether or not Earthlings, utilizing their wits and entrepreneurial spirit, might find ways to survive in outer space.[2]

In mainstream Western culture, it is often assumed that Indigenous artists are uninterested in science fiction or that they are unconcerned about the cosmos or cosmic travel, because they are more concerned with planet Earth, or "Mother Earth" (a term often associated with Indigenous cultures). But these kinds of stereotypical assumptions are far from true. As one example, we can consider the short story by Simon Ortiz (Acoma Pueblo), "Men on the Moon" (which has since been reprinted many times, including in Ortiz (1999). Ortiz, the author of the Foreword for this volume, is one of the most honored writers in the global Indigenous arts community and helped catalyze what is often called the post-1960s "Native American Literary Renaissance" that began with the publication of N. Scott Momaday's *House Made of Dawn* (1968). In Ortiz's short story, an Acoma grandfather, Faustin, receives his first television from his grandchildren, who want him to see the Apollo mission, where American astronauts will land on the moon for the first time. The astronauts, his children tell him, are going to the moon to collect rocks and to bring them back to Earth. Why, Faustin asks, do men want to collect rocks on the moon? Because, his children and grandchildren tell him, they are trying to find knowledge to better mankind and learn more about the universe in which we live (Ortiz 1999, 11).

At first Faustin thinks that his grandchildren must surely be joking. Hasn't "anyone ever told them" where and how "everything began?," he asks incredulously (Ortiz 1999, 12). Other people have "tried to tell them," the children say, but "they want to find out for themselves" (Ortiz 1999, 12). Faustin chuckles after hearing this answer. That night, Faustin dreams about two characters from ancient Acoma oral traditions, Flintwing Boy and Coyote. In Acoma culture, Flintwing Boy is associated with the constellation that the ancient Greeks called Orion, and both Acoman and the Greek associate with the hunter. Looking out over the hills near Acoma, Flintwing Boy sees a giant machine land on the Earth. As it begins moving, its giant metal legs crush all the trees, flowers, and grass in its path, scaring away the deer (Ortiz 1999, 11). Coyote is frightened. Flintwing Boy calms his friend by taking corn pollen ("corn food") from the bag hanging from his side and placing the pollen in several places on Coyote's body. With the "corn food" still in his hand, Flintwing Boy faces east, where the sun rises every morning. Praying, he says, "We humble ourselves again. We look in your direction for guidance. We ask for your protection. We humble our poor bodies and spirits because only you are the power and the source and the knowledge" (Ortiz, 1999, 9).

Ortiz's use of the words "you," "power," "source," and "knowledge" in the context of the short story's humorous look at NASA space missions evokes key concepts in Indigenous understandings of Earth as a "you"with relations to the larger cosmos: a "person," or a "sentient" or "creator" being. The dusty yellow corn pollen in Flintwing Boy's hand is understood, literally, as an earthly manifestation of the sun's power. Corn as a plant turns sunlight and carbon into the pollen needed to fertilize each "egg/seed" of corn, which is both an important food and an important symbol in Acoma culture. Therefore, pollen is understood literally as a source of power and life. Faustin's dream is embedded within sophisticated scientific understanding of cosmic and biospheric sources of power and knowledge upon which humans depend for their sustenance and well-being. This knowledge is central to emerging Western understandings of multi-natural worlds and pluri-versal, rather than universal, understandings of the cosmos. Faustin's dream points to one of the overarching themes of this collection, Indigenous "cosmovisions," while Ortiz's story concisely reveals the reasons why, from the Arctic North to the Antarctic South, Indigenous understandings of "you," or "sentient creator beings," suggest a cosmos of relations that speak to complex entanglements of the human with the more-than-human that must be creatively and thoughtfully negotiated.

Such negotiation stands in contrast to the imagery of the giant metal monster in Faustin's dream. This monster denotes a careless, reckless disregard for the pluri-verse and serves as an implicit critique of EuroAmerican understandings of power as colonization of frontiers, wilderness, and outer space. The concepts of frontier and wilderness have undergirded the colonial enterprise, and have justified the occupation of "empty" continents. These concepts, beginning in the early 1400s, were invoked to authorize genocide and slavery in the Americas and Africa on a grand scale. As M. Jane Young (1987) observed at a convocation of Native American scholars in the 1970s, at a time when NASA was training astronauts in Monument Valley on the Navajo Nation, a group of Native Americans was asked about their attitudes toward space travel. Their response was that they would "pity the poor buffalo and Indians of outer space," by which they proactively evoked the history of death and chaos that accompanied notions of Western civilization and progress (271). This response to questions about U.S. space travel elucidates Faustin's chuckle after his grandchildren tell him that the mission to the moon will result in "knowledge." If, Faustin concludes, these men on Earth are using their machines to crush trees, flowers, and grass, but then going to the moon in search of the "tiniest bit of life," his new television will be better used to watch a wrestling match on another channel between an Apache and an American than the Apollo moon mission. Faustin hopes the wrestling match will be won not by a settler colonial but by the Apache (Ortiz 1999, 14).

From Flintwing Boy's place in the night sky to the Navajo conception of Father Sky and Mother Earth, to the Mayan conception of Sea and Sky, recorded in the *Popol Vuh* (a corpus of mytho-historical-astronomical narratives), Indigenous groups around the world see themselves as intimately related to the sun, moon, stars, earth, and water. In the *Popol Vuh,* the gods of the sea and sky make four attempts to create human beings. In one failed attempt, they create a being of mud that is washed away by the rain. Mayan elders suggest that this story is not a paraphrase of the Greek story of Prometheus or the biblical story of Adam's creation from mud, but a "negation of Adam," and probably a direct critique of the beliefs and practices forced on Quichéan peoples by Spanish colonizers who did not believe the Maya to be human (Tedlock 1985, 263–64, 270–71). Joni Adamson has suggested the phrase "seeing instrument," long associated with the *Popol Vuh,* to describe cultural productions and humanities scholarship that draw our attention to Indigenous "cosmovisions"—conceptions of entangled human relations with more-than-human worlds (Adamson 2001, 145; Adamson 2014). From ancient Indigenous story cycles and farmers' almanacs to contemporary novels, such narratives have long worked as an imaginative force for thinking about "the origins and [ongoing evolutionary] transformations of the world and its inhabitants" (Cruickshank 2005, 99).

This is the reason why—well ahead of the Paris Climate talks and inspired by "cosmovisions"—many representatives from the Global South and myriad Indigenous groups convened in Cochabamba, Bolivia, in 2010 for the "World People's Conference on Climate Change and the Rights of

Mother Earth." There they wrote a Universal Declaration on the Rights of Mother Earth (UDRME, 2010). The working papers from this conference show the participants' intention to attribute to Indigenous peoples, nations, and organizations ancestral "cosmovisions"—thousands of years in the making—that conceive of Earth as a "person" or "living being with whom [all persons] have an indivisible and interdependent relationship" ("World Peoples' Conference" 2010, Common Objective 1).[3] The Working Group 7 on Indigenous Peoples decried "aggression toward Mother Earth" as an assault on us," with the word "us" clearly meant to convey all human groups (not just Indigenous) *and* all other "persons," including the "soils, air, forests, rivers, [and] lakes" ("Final Conclusions" 2010, parag. 2).

The UDRME notion of all living, mutually interdependent entities as "persons" that draw on ancient traditions is at the center of *Ecocriticism and Indigenous Studies*'s focus on cosmovisions. Like Flintwing Boy and Coyote, and Father Sky and Mother Earth, the chapters of this volume reveal how many Indigenous cultural traditions throughout the world imbue their worlds with agential "persons." "Whether human or animal in form or name, these characters [behave] like people, though many of their activities are depicted in a spatiotemporal framework of cosmic, rather than mundane, dimensions" (Adamson Clarke 1992, 45, note 11).[4] When represented in contemporary poetry, fiction, or film, they work in a time-space continuum of past, present, and future, because, like Flintwing Boy, they simultaneously comment on the "relic" or ancient story, and the new genre (such as "Men on the Moon's" written short story form), to help speculate about the present and the future in a rapidly changing contemporary world. Flintwing Boy, then, is a seeing instrument employed by Faustin to think about what he is seeing on the television, assess it, put it into the context of what the Acoma people have learned through time, and consider how this new information might be useful in the present and future.

Anthropologists Eduardo Viveiros de Castro (1998, 2004) and Marisol de la Cadena (2010), in the contexts of their work with Indigenous peoples in Brazil, Bolivia, and Ecuador, have written about "transformational entities," or "persons" that offer insights into the "multiverse." The world in which humans live, according to stories about persons such as Flintwing Boy (and the many other beings encountered in this volume) that provide insight into a multi-scaled world of plants, animals, and microfauna (and even sacred mountain, lakes, or rivers) that are considered "beings," are all connected and entangled together. Such entities were the subject of a 2015 conference organized by Viveiros de Castro, *Os Mil Nomes de Gaia* / Thousand Names of Gaia, where Donna Haraway explained how "persons" from the oral traditions of Indigenous groups often name "collected things."[5] From the macro-scaled planetary body of Mother Earth to the micro-scaled fungal networks of a Brazilian rainforest that might be named *Sachamama*, meaning "Mother of Forests," they give presence to myriad entangled temporalities and spatialities, or "entitities-in-assemblages—including more-than-human, other-than-human, inhuman" (Haraway 2015, 160).

Ecocriticism and Indigenous Studies takes the pulse of current Indigenous artistic diversity and political expression, which work as seeing instruments and speak of "cosmovisions," "persons," and "collected things." In so doing, this volume places itself squarely at the intersection of the environmental humanities and Indigenous Studies. Both interdisciplinary areas of study have seen remarkable scholarly growth in the last twenty years, with distinct yet often parallel concerns that this collection works to capture. In fact, as we will show, the environmental humanities broadly, and ecocriticism and Indigenous studies specifically, emerge out of the same long, entangled, historical roots. Both can be traced, as we show in the next section of this introduction, to global Indigenous oral narrative archives, or stories about "persons" or "collected things."

The Histories of Cosmovisions, the Roots of Indigenous Studies, and Ecocriticism

Ecocriticism and Indigenous Studies furthers conversations begun almost fifteen years ago in Joni Adamson's *American Indian Literature, Environmental Justice and Ecocriticism: The Middle Place* (2001). This book was the first ecocritical monograph to articulate the importance of oral and written Indigenous literatures that were confronting EuroAmerican conceptions of "nature" and "place." Adamson drew the attention of ecocritics to issues of environmental justice and how constructs such as race, class, and gender cannot be separated from concepts of "nature." *American Indian Literature* helped propel ecocriticism's twenty-first century turn—its expansion from a purview of EuroAmerican texts of nonfiction nature-writing—to the inclusion of the voices and struggles of marginalized minorities and specifically to the struggles of global Indigenous communities engaging in innovative coalitional work. Adamson's work was enriched and expanded by other ecocritics also working on these issues. Patrick Murphy's *Further Afield in the Study of Nature-Oriented Literature* (2000), and Adamson's co-edited collection (with Mei Mei Evans and Rachel Stein), *The Environmental Justice Reader: Poetics, Politics, and Pedagogy* (2002) are only two examples of what is now a considerable canon of ecocritical work—from numerous journal articles to anthologies such as Graham Huggan and Helen Tiffin's *Postcolonial Ecocriticism: Literature, Animals, and Environments* (2010); Elizabeth DeLoughrey and George Handley's co-edited *Postcolonial Ecologies: Literatures of the Environment* (2010); Joni Adamson and Kimberly Ruffin's *American Studies, Ecocriticism, and Citizenship: Thinking and Acting in the Local and Global Commons* (2013); and Serenella Iovino and Serpil Opperman's *Material Ecocriticism* (2014). There are also myriad monographs, such as Deborah Bird Rose's *Dingo Makes Us Human* (2009) and *Wild Dog Dreaming: Love and Extinction* (2011); Rob Nixon's *Slow*

Violence and the Environmentalism of the Poor (2011), and Sarah Jaquette Ray's *The Ecological Other: Environmental Exclusion in American Culture* (2013). Each of these works illustrate the powerful directions ecocriticism—and the environmental humanities—are taking in the twenty-first century to engage environmental justice, postcolonial ecology, multispecies ontologies, and Indigenous studies.

Adamson's *American Indian Literature* began tracing the entangled roots of both Indigenous studies and ecocriticism in oral archives of cosmological narratives told from North to South America, the Caribbean, and Africa. These stories began to be systematically collected in the late nineteenth and early twentieth century by the Bureau of American Ethnography and were published as reports (still found in many U.S. university libraries). Numerous tribal North American groups, and the anthropologists and ethnographers contributing to them, consulted these reports throughout the twentieth and twenty-first centuries to argue for land rights claims in court. At the same time, renowned Native North American Indigenous writers such as N. Scott Momaday (Kiowa), Louise Erdrich (Anishnaabe), and Leslie Marmon Silko (Laguna Pueblo) were also consulting Bureau of American Ethnography reports on their own tribal groups and using the stories and information they found there to piece together cultural and ecological knowledge that had been scattered or lost over years of systematic oppression and racism. Piecing this information together with the stories they could still hear from their living relatives, they wove "persons" into their contemporary poetry and novels, which laid the foundation for the "Native American Renaissance" (Adamson 2001, 175–76).

The Bureau of American Ethnography reports have a genealogy that extends back to German geographer Alexander von Humboldt's travels in the Amazon at the end of the eighteenth century and, more specifically, to his interactions with the first peoples of the Americas that he met there. It is here, in Humboldt's travels, that we begin to see the entangled roots of Indigenous studies and ecocriticism. Laura Dassow Walls describes how Humboldt's wide-ranging research in Europe and South America influenced an intellectual network that included figures—both mainstream and dissident—who would deploy his ideas throughout the nineteenth century and into the twenty-first (2009, 268). One of the "dissidents" was Franz Boas, a German anthropologist, who studied the Inuit communities of what is now considered the Canadian North (specifically Baffin Island) and British Columbia. Calling himself a "cosmographer," Boas considered every phenomenon "worthy of being studied for its own sake" and, at Columbia University, trained scholars who would shape not only the field of ethnography but also budding arts movements including the Harlem Renaissance. Some of these scholars were Edward Sapir, Ruth Benedict, and Zora Neale Hurston[6] (Adamson 2016; Rose 2015, 110; Walls 2009, 117, 211). Humboldt's friendship with Simon Bolivar also influenced the latter's move to fight for Latin America's autonomy from Spain (Walls 17, 97–98). A "cosmographer," Humboldt told his students, "studies the history of phenomena, what they are and how they came to be just that way" and "cherishes the very particulars that science uses then throws away" (Walls 2009, 212).[7]

Walls traces Humboldt's resurrection of the word "cosmos" from the archaic Greek for the title of his bestselling five-volume work, *Kosmos*, which influenced a generation of thinkers on several continents (Walls 2015, 48). Walls argues that we need to recover the ancient uses and understandings of "cosmos," in Indigenous thought, in ancient Greek thought, and from Humboldt's uses of the word, since popularized contemporary use as a scientific object or Christian religious symbology often "masks its long history as humanity's oldest ecological vision of our planet" (Walls 2015, 47).

At this point, it is especially important to note that Indigenous cosmovisions influenced Humboldt rather than the other way around. His encounters with Indigenous peoples and their interactions with "persons" can now be seen, through Boas, to have shaped anthropology, ethnography, and later, what we now refer to as Indigenous studies. Humboldt's encounters with Indigenous belief systems in the Amazon also influenced a generation of thinkers today (recognized as the founders of American environmentalism and the inspirations for ecocriticism), including Ralph Waldo Emerson and Henry David Thoreau (Walls 2009, 268). Humboldt defined "nature" as "a planetary interactive causal network operating across multiple-scale levels, temporal and spatial" (Walls 2009, 11), an understanding we see, for example, clearly at work in Thoreau's *The Maine Woods* (1864). Today, with these deep histories in mind, early 1990s ecocritical perceptions are being reviewed, rearticulated, and reconsidered in ways coming to be called, both inside and outside academia, "cosmopolitics."

"Cosmopolitics" implies that we are entering a moment in politics that takes as its goal, to use the words of Ecuador's recently revised constitution, the recognition of intergenerational, evolutionary space and time required not just for the survival of all species, but for the recognition of the "rights" to life for all humans and nonhumans (Adamson 2014; Ecuador Constitution 2008). Influenced by his friendship with Eduardo Vivieros de Castro and the long histories of the word "cosmos" outlined above, Bruno Latour (2004) spotlights its Greek meaning of "arrangement" and "harmony." He argues for the word as a synonym for "the common good world"—that is, a world that brings together, by "cosmopolitics" or "due process," a pluriverse of peoples and natures (rather than a universe,) living in "a commons resilient enough to embrace the future" (Latour 2004, 237–40; Walls 2015, 49).

Similarly, Isabelle Stengers, in her path-breaking essay, "The Cosmopolitical Proposal", cautions us to "slow down" (2005, 995), lest we, to use the words of Laura Dassow Walls, impose our "particular cosmos or tradition upon all" (Walls 2015, 49). Stengers extracts from the word "cosmopolitan" its two constituents: cosmos and politics. "Cosmos" refers to the unknown, or what we generally still do not know about multiple, divergent worlds, and "politics" to the articulation of the unknown, which we "would eventually be capable" of articulating (Stengers 2005, 995). She then argues that a cosmos detached from politics is irrelevant and dives deep into the philosophies of politics and science to explore how "our modern world," to use Bruno Latour's phrase, separated humans from nature (Latour 1993, 27; Stengers 2005). Despite agreeing with Latour's and Stengers' cosmopolitics agenda to redress binaries—of human/nonhuman, culture/nature, science/social—some scholars take issue with such postmodern theories of cosmopolitics, cautioning against their abstract quality (e.g., Chagani, 2014; Cubitt 2014; Shapiro, 2005). In effect, as ecomedia scholar Sean Cubitt notes, theories of environmental criticism "require[s] both a political and an ethical stance, one in particular that ponders the likely outcomes of a course of action" (2014, 167).

Cosmopolitics in the context of Indigenous struggle does exactly this. As seen in anthropologist Marisol de la Cadena's situating of "cosmos" in Latin America political alliances that include Indigenous communities and activists, Indigenous epistemologies and cosmovisions cannot be unraveled from politics or ethics (2010). Nowhere is this more clear than in its evocation in the UDRME, which centers multispecies relations in the context of climate change and the eco-social injustices that accompany its unfolding. Highlighting contemporary case studies of Indigenous struggle, the chapters of *Ecocriticism and Indigenous Studies* further demonstrate that because cosmovisions are articulated within specific histories, geographies, and contemporary contexts, they are an active means to negotiate the practice of daily survival. They serve as philosophical engagements to navigate the everyday ethics of living in wider worlds with humans and nonhumans alike.

To recognize Indigenous cosmovisions as participating in everyday and situated practices is to also comprehend them as dynamic epistemologies. They are always in the process of being interpreted. Thus, it is imperative to understand that cosmovisions are not essentialist or simplistic answers to ecological crises. Grounded in context, they can be limiting or liberating in their ethics and their politics as applied to other "persons." For example, as Adamson (2013) discusses when analyzing Valliere Richard Auzenne's and Juan Carlos Galeano's documentary, *The Trees Have a Mother* (2008), Auzenne and Galeano show how some individuals living in the Amazon engage in the practice of cutting the penises off dolphins who are accidently caught or drowned in fishing nets, drying the penises, and selling them to individuals hoping to boost their human (male) virility, a practice some justify based on stories about dolphin virility (Adamson 2013). In addressing the ethics of such practices, Galeano does not shy away from presenting the everyday contexts in which such traditional stories are engaged (Galeano 2009, 19–38). As he shows in all his work, people living in the Amazon basin must work for their eco-cultural livelihoods. He shows that opportunities for cultural and economic well-being have been drastically eroded. With mega-corporation trawlers engaged in overfishing, wildcat gold mining polluting streams, and deforestation intensifying erosion, gross social disparities and ecological resource devastation are becoming more widespread. Yet, practices surrounding dolphins, as referenced above, caution us against considering Indigenous stories or cosmovisions a panacea for ecological viability and remind us of the complex situatedness of local practices.

However, in many cases, cosmovisions articulate dynamic epistemologies that have been negotiated over long histories (sometimes thousands of years), and many present sound ethical and scientific reasoning for ecological protection. Passed down through the centuries, they offer models of long-term conservation practice and scientific knowledge. Indigenous practices have often been sidelined as "superstition" by those who consider mainstream Western sciences "objective" and superior. However, in collaboration with Indigenous communities, ethnographers and Indigenous studies scholars have continued to collect the oral astronomical, ceremonial, cultural, agro-ecological, and ethno-botanical knowledges of diverse ethnic groups around the world and consider them legitimate and scientifically sound. Such oral knowledges are understood more and more as "archives," or sophisticated "cosmographies" (Walls 2009, 212), that have allowed communities to express their own "scientific" literacies about ecosystem connections. It is from these oral archives that Humboldt expanded his European-based knowledge all those years ago when he encountered Amazonian Indigenous communities. Today, such expanded understandings are the foundation of research understood as "multispecies ethnography" or "multispecies relations" (Galeano 2009; Kirksey 2010; Rose 2015). Inherently interdisciplinary, multispecies ethnography draws together natural scientists, social scientists, environmental humanists, and artists who acknowledge Indigenous cosmovisions as insight into the eco-cultural entanglements of human with nonhuman and seek to learn about the "host of organisms whose lives and deaths are linked to human social worlds" (Kirksey 2010, 545).

Expanding on this growing interdisciplinary recognition of Indigenous eco-agency, *Ecocriticism and Indigenous Studies* turns its attention to Indigenous cosmovisions as the practice of multispecies relations in contemporary, twenty-first century contexts. Through case studies from around the world, it grounds such relations in the politics and ethics of everyday practice, which for Indigenous communities is often the struggle for eco-cultural survival at the frontlines of climate change and escalating ecological destruction. Through myriad expressions of cosmovisions, *Ecocriticism and Indigenous Studies* also illuminates the dynamic aesthetics of such practices. Its contributors engage with various storytelling modes—from literature, poetry, and film to opera, performance, landscape art, and even the land itself. Situating discussion of these various media in one comprehensive collection, the volume innovatively engages the expanding media focus of twenty-first century eco-activism and encourages productive intertextual conversations necessary to interdisciplinary thinking that is required to confront the cosmic scope of unprecedented planetary change in the twenty-first century.

Taking the Pulse of the Present: The Chapters

Ecocriticism and Indigenous Studies is about collaborations. Its conversations "from Earth to Cosmos" bring together scholars and artists, Indigenous and non-Indigenous partners, and ecocritics and Indigenous activists in a volume organized on three key tenets: resilience, resistance, and multispecies relations. Central to Indigenous eco-activism, these concepts serve as important reminders of this volume's political and ethical foci. Specifically, *resilience* articulates ongoing Indigenous responses to centuries of politically enforced extermination, assimilation, and marginalization; *resistance* highlights active struggles for self-determination and sovereignty against cultural and eco-genocide; and *multispecies relations* illuminate the philosophies that undergird Indigenous ecological literacies often applied in the practice of resilience and resistance.

While these tenets frame the volume, as each chapter illuminates, resilience, resistance, and multispecies relations cannot be easily disentangled from each other. Instead, together they express the complexities of Indigenous activism around cultural and scientific understandings of "environment." These complexities are revealed through interdisciplinary intersections that link Indigenous studies and ecocriticism to critical methods in research areas such as film studies, science studies, performance studies, history, anthropology, and new media studies. Contextualized within these fields, the central theoretical concept of cosmovisions is paired with important emerging ideas—material ecocriticism, Indigenous ecocinema, environmental governance, affect, and biosemiotics—thus, showcasing new scholarly directions in the environmental humanities. At the same time, the volume grounds itself firmly in artistic practice. Along with close theoretical analyses of Indigenous-authored works, *Ecocriticism and Indigenous Studies* includes chapters that are creative reflections by artists themselves. Thus, the book works actively to destabilize rigid notions of academic work and to remind its readers of the importance of storytelling as decolonizing practice. As Indigenous literary scholar Mishuana Goeman writes, "the imaginative possibilities and creations provided in the play of a poem, imagery of a novel, or complex relationships set up in the short story provide avenues beyond recovery of a violent history of erasure and provide imaginative modes to unsettle settler space" (2013, 2). Her words can be extended to the film, opera, poetry, and paintings of the artist-authors present in this collection.

In its broad scope, *Ecocriticism and Indigenous Studies* draws on a talented and diverse group of contributors. The first chapter of *Part I: Resilience*, "Negotiating the Ontological Gap" features the collaborative analysis of Aotearoa/New Zealand-based scholars and artists, Janine Randerson and Amanda Yates. They showcase how the video installation, performance art, and landscape architecture of artists Natalie Robertson, Shannon Te Ao, and Amanda Yates herself engage the Māori concept of *whakapapa*, an ontology of connection where humans are situated within an extensive and inclusive geneology. Articulating how such cosmologies highlight resilience in the face of colonial destruction, they argue that Māori art and design function as eco-political practices that activate the "ontological gap" (Hunt 2014)—a creative, negotiable space where worlds can meet.

Contemporary Indigenous film practice similarly demonstrates such "ontological gaps" as seen in Chapter 2, "Science Fiction, Westerns, and the Vital Cosmo-ethics of *The 6th World*," by ecocinema scholar, Salma Monani. Drawing on emerging scholarship about Indigenous speculative fiction (Dillon 2012, Lempert 2014), Monani demonstrates how Navajo filmmaker, Nanobah Becker's short space travel film, *The 6th World: An Origin Story,* critiques colonial and neocolonial histories to help us re-think film genre, Indigenous continuance, and ecological relations. Specifically, she suggests that it visualizes a "vital cosmo-ethics" that argues for humans as one amongst many in a complex, pluriverse that must be respectfully and creatively negotiated. While an engagement with *The 6th World* embraces "Indigenous futurisms," Chapter 3, "Long Environmentalism: After the Listening Session," provides an explicit focus on historical process to articulate Indigenous resilience. Artist and public intellectual Subhankar Banerjee engages the intergenerational eco-social struggles of Alaskan Iñupiat and Gwich'in communities against corporate resource extraction to introduce the concept of "long environmentalism." Illuminating his own decade-long involvement in these struggles, he explains that the two principal tenets of long environmentalism include (a) collaboration among unlikely allies through the act of sincere listening; and (b) a period of time that is long enough to enable what was once considered marginal to become significant. Banerjee argues that while holding varieties of contradictions in place, long environmentalism is historically grounded, exposing past injustices. It is also futuristic, enabling radical hope for sustainable human and nonhuman biotic life.

In the fourth and final chapter of *Part I*, Navajo Nation poet laureate Laura Tohe (Diné) similarly reaches into the past to articulate Indigenous resilience into the present. Her chapter reflects on the creative and collaborative process involved in the making of a contemporary operatic performance, *Enemy Slayer, A Navajo Oratorio* (2008), a libretto, which she authored. As Tohe explains, the oratorio, commissioned by the Phoenix Symphony Orchestra became a response to veterans damaged by post-traumatic stress disorder. The story is grounded within the ancestral space, or Diné *bikeyah,* which forms the Navajo notion of homeland and their cosmovisions of earth and sky. The returned war veteran, represented by the mythic figure Seeker, is restored to health through engagement with his ancestral geographic and spiritual homelands.

In *Part II: Resistance*, contributors are equally interested in articulating how past traditions influence present circumstances in complex ways that contribute to an engaged politics of resistance. In Chapter 5, "Dancing at the End of the World," ecocritic Janet Fiskio draws on Judith Butler's work on precarity and Susan Leigh Foster's performance theory of "the choreography of protest," to explore the politics and poetics of bodies in "stillness" and "movement" in on-the-ground demonstrations of Indigenous social movements like Idle No More and the Healing Walk. Fiskio shows how, with their particular focus on anti-extraction and anti-pipeline activism, such corporeal demonstrations intervene in public and sacred spaces to reveal the violence of colonialism and global capitalism even as they articulate a vision of sovereign communities. While Fiskio's attention is specifically on the presence of bodies on-the-ground, in Chapter 6, "New Media, Activism, and Indigenous Environmental Governance," ethnic studies scholars Clint Carroll (Cherokee) and Angelica Lawson (Northern Arapaho) turn our attention to new media expressions of resistance. They examine how the Ojibwe website maiingan.org articulated opposition to state-sanctioned wolf hunts in Minnesota and Wisconsin. They argue that the website expressed a uniquely Indigenous "media cosmology" (Loft and Swanson, 2014). To contextualize such cosmology, they situate their website analysis within the framework of Indigenous "environmental governance"—specifically, how American Indian nations struggle to assert their traditional perspectives toward nonhuman beings as *relatives* against a dominant resource management paradigm, which sees nonhuman beings in instrumental and hierarchical ways.

In Chapter 7, "Cyclical Conceptualizations of Time," Finnish ecomedia scholar Pietari Kääpä likewise highlights resistance in screen media, returning our attention to cinema. Kääpä traces the history of Sámi cinematic representation in Nordic countries in conjunction with commentary on regional development schemes to articulate the complex eco-politics of Sámi recognition. Central to his argument are Sámi cinema's distinct notions of "ecological temporality" (as cyclical and of long-time) and "banal everydayness" (a mode of seemingly depoliticized representation that concerns the politics of performing the everyday). Such notions, he demonstrates, present challenges to Sámi filmmakers' in Western industry contexts even as their pluriversal aesthetics help visualize alternative ecological modes to hegemonic notions of progress and stereotypical ideas of Sámi as exotic other.

In Chapter 8, "Resistance and Hope in Mohawk Cinema," Shelley Niro and Salma Monani similarly undermine Indigenous stereotypes in film while drawing attention to the cosmopolitics of Niro's visions. Bracketed by an introduction and conclusion by Monani, Niro provides a retrospect on a lifetime of her filmmaking practices. Prefaced by memories of her childhood, Niro reflects on her cinema. The collaborative chapter demonstrates how Niro's Mohawk sensibilities generate a richly complex cinematic space that weaves in Iroquois cosmologies to prompt her viewers, both Indigenous and non-Indigenous, to encounter a world which questions binaries—of past/present, material/ethereal, self/other—and as a consequence of the very ideas of nature/culture.

Interrogating such Western binaries become even more apparent in the chapters of *Part III: Multispecies Relations*, where the "personhood" of other beings and their ecocultural entanglements with human lives come into explicit focus. In Chapter 9, "Network of Networks: Multispecies

Stories," Kathryn Yalan Chang considers how Linda Hogan's (Chickasaw) 1995 novel *Solar Storms*, which fictionalizes hydroelectric impacts on Indigenous communities, anticipates filmmaker Joel Heath's 2011 documentary, *People of a Feather*. Chang uses cultural anthropologist Arjun Appadurai's phrase "globalization from above" (2001) to argue that dam building disrupts Indigenous peoples' materially situated links to their homelands. Yet, she also suggests that both texts resist such disruption through their attention to all things as "storied matter" (Iovino and Oppermann 2014) and to "trans-Indigeniety" (Allen 2012). Together these concepts foreground multispecies relations and show how material engagements can fuel community solidarity to enable what Chang terms a "network of networks." Unlike in Canada, where Chang focuses her analysis, in India Indigenous peoples lack networks of trans-Indigenous support. Yet, as ecocritic and filmmaker, Rayson K. Alex's chapter, "*tinai*-Documentation as Ecocultural Ethnography" (Chapter 10) suggests, they are nonetheless strongly grounded in networks of multispecies relations, which fuel contemporary localized struggle. Alex's ethnographic reflection of his participation in the Mudugar tribal community of South India frames Mudugar cosmovisions within the local concept of *tinai*, a geo-poetic convention and social order specific to the early Dravidian peoples of Southern India. He presents four stories that suggest how multispecies relations are evoked as ecocultural rootedness. In turn, this rootedness serves to ground Mudugar resistance against long legacies of marginalization by settler cultures. Alex argues that in the face of a current onslaught of rapid, capitalist-driven development in India, not paying attention to the Mudugar and their cosmovisions is inherently a grave environmental injustice; *tinai*-documentation attempts to address this injustice. In effect, Alex makes a concerted effort to energize a collaborative vision that can assist India's Indigenous struggles develop the beginnings of what Chang (Chapter 9) is calling a "network of networks."

Abigail Pérez Aguilera's "The Tangibility of Maize" (Chapter 11) turns our attention to Mexico and the figure of corn as a motif in the bilingual literary work of Mikeas Sánchez's (Zoque from Chiapas) *Mojk'jäyä-Mokaya* (2013) and the bioart installations of Minerva Hernández Trejo and Héctor Cruz. By engaging corn as an agential being, all these artists, she argues, critique the violence of neoliberal systems that threaten local food sovereignty. Sánchez's poetry draws on her Zoque cosmologies of corn to resist heteronormative and gendered violence towards women, while Minerva Hernández Trejo and Héctor Cruz's bioart highlight corn as "alive" to prompt more comprehensive understandings of the entangled relations between human and nonhuman. Together the works foreground "Indigenous knowledge systems" (IKS) as resistance against debilitating neoliberal control. The final chapter in the collection, "Why Bears, *Yakumama* (Mother of All Water Beings), and Other Transformational Beings Are (Still) Good to Think," similarly illuminates other-than-humans as agential beings. Co-authored by Joni Adamson and Juan Carlos Galeano, with vivid illustrations by Peruvian artist Solmi Angarita, it captures the rich oral traditions of Amazonian Indigenous peoples that are still powerfully at work in modern communities today. Building on the ecocritical scholarship of Joni Adamson, the chapter explores how Galeano's poems (and Angarita's paintings) present transformational beings that are "good to think" and why so many convey a humorous tone that suggests not only feelings of optimism and hope for the future, but a marked focus on building livable worlds in which health and well-being are "cosmopolitical" goals. The chapter also reflects on Galeano's work to pilot waste management solutions in the small town of Padre Cocha in Peru. Galeano's project is linked to "Humanities for the Environment" projects being piloted by Adamson and others who are interested in exploring how stories can be employed to create desirable future worlds.

Expanding Horizons, Cosmic Futures

In *Walking the Clouds: An Anthology of Indigenous Science Fiction* (2012), Grace L. Dillon points out that Indigenous communities have long understood themselves to be living in a "post-Native Apocalypse" (Dillon 2012, 10). In response to this "apocalypse," many Indigenous artists creatively imagine worlds that have qualities of the "speculative arts," or art that explores "Indigenous futurisms" where characters or "persons" help communities recover and survive for viable, healthy futures (ibid). *Ecocriticism and Indigenous Studies,* with its focus on multiple artistic responses to contemporary ecological crises, draws our attention to such speculations in practice. Central to the speculations presented here are cosmovisions. And while cosmovisions draw on "the past," they confront the politics and ethics of contemporary and future eco-crisis to articulate a sophisticated understanding of "science fiction."

Specifically, such engagements usually spotlight "Indigenous scientific literacies" as complex, multispecies entanglements that imaginatively argue for a safer, livable present and future. As Dillon writes, such "science fiction" is distinct from somewhat limited understandings of the genre that focus on modern or future technologies as the "science" that saves us (2012, 3). Instead of a technological fix to current problems, the "science fiction" articulated by the essays in *Ecocriticism and Indigenous Studies'* insist that lasting solutions must be rooted in clear-sighted understandings of multi-faceted human/more-than-human relationships that exist in complex amalgamations of a "space-time continuum." Whether through grandfather Faustin's dream world of Flintwing Boy and Coyote in Ortiz's *Men on the Moon* (1999), astronaut Redhouse's ceremonial evocation of Changing Woman in Nanobah Becker's space travel short film, *The 6th World* (2012), the operatic songs of Seeker in Laura Tohe's libretto for *Enemy Slayer*, or the rhythmic round dances of Idle no More demonstrations, in these case studies, one sees "pasts, presents, and futures that flow together like currents in a navigable stream" (Dillon 2012, 3). Such "native slipstream" or "space-time thinking" (Dillon 2012, 7–8) is particularly valuable as a means to confront Western notions of linear progress, which not only tend to ignore the cyclic rhythms of earth's bio-geomorphic systems but do so with dangerously myopic focus.

The short-sighted hubris of notions of "progress" are becoming more apparent through globally publicized events like the United Nations' 2015 Conference on Climate Change in Paris. The chapters in *Ecocriticism and Indigenous Studies* draw our attention to many equally important though often less publicized struggles against such hubris. In highlighting how Indigenous communities from across the world draw their ancient cosmovisions into the "slipstream" of the present, they demonstrate what Indigenous scholar, Dean Rader calls "engaged resistance"—a fundamentally Indigenous form of aesthetic discourse that engages both Native and Western means as resistance against "colonial assimilation and erasure" and toward "survivance" (2011, 1). Using twentieth and twenty-first century technologies like cinema and new media, older written expressions of literature and poetry, age-old ceremonial dance and song, and markings on the land itself, Indigenous artists and their communities revitalize storytelling modes that engage the human with the more-than-human, and draw into view earthly and cosmic "persons" as allies in struggles against localized and planetary destruction. Such struggle is not only focused on critique and resistance but on continuance.

As many of the chapters in *Ecocriticism and Indigenous Studies* indicate, these struggles are not easy, yet they are also examples of how cosmovisions play a powerful role in helping tilt the political balance towards more ethical considerations of wider worlds. From the stay on construction of the XL Keystone pipeline, which puts pressure on wanton oil exploration in Canada's First Nations' territories, to a reconsideration of the removal of wolves from Minnesota and Wisconsin's endangered species list, from the withdrawal of oil and gas mining permits in the biodiverse and fragile Alaskan Arctic, to victories for food sovereignty, the chapters in this volume evoke the possibilities of engaged resistance. As the chapters also make clear, many of these political victories are tenuous. Many were being negotiated even as the authors were writing their chapters, and by the time this volume is published and being read, these victories might again be in jeopardy. Yet, they speak to what Banerjee (in Chapter 3) suggests is a sense of radical hope.

By offering readers these case studies at the intersections of ecocriticism and Indigenous studies *Ecocriticism and Indigenous Studies* invites its readers to radical hope, where we might better imagine our pasts, presents, and futures as ethically answerable to other "persons" in a locally situated, globally networked, and cosmologically complex world.

Notes

1. For a discussion written by members of the Anthropocene Working Group on the current status of this term in the scientific community, see Zalasiewicz, Williams, and Waters (2015).
2. Another blockbuster film, *Interstellar* (2014), has a similar "space as humanity's possible redemption" plot. While these films use space as an escape frontier, a host of other recent films use climate change on Earth as a speculative device against which their plots enfold (e.g., *Mad Max-Fury Road* 2015; *The Last Survivors* 2015; *Snowpiercer* 2013; and even the campy *Sharknado* 2013).
3. See also the thinking of Vivieros de Castro (1998, 469).
4. For further elaboration on why transformational animals or persons are "good to think," see Babcock (1975).
5. For more information on the conference, see https://osmilnomesdegaia.eco.br.
6. Hurston based her novel, *Their Eyes Were Watching God* (1937), on her fieldwork in Florida and the Caribbean, conducted in the "participant observer" tradition of Boas, which was the major methodological innovation of this field at the time.
7. Ironically, Humboldt's fame was eclipsed by his own disciple, Charles Darwin, who carried to the next level Humboldt's spatial and evolutionary vision. However, faced with an evolutionary "abyss" without a basis in Anglo-American religious beliefs, Darwin's solution, as Dipesh Chakrabarty has argued, was to delink the humanities from the sciences and to "treat the environment simply as a silent and passive backdrop" to historical narrative (Chakrabarty 2009).

References

Adamson, Joni. 2001. *American Indian Literature, Environmental Justice, and Ecocriticism: The Middle Place*. Tucson: University of Arizona Press.

———. 2013. "Environmental Justice, Cosmopolitics and Climate Change." In *The Cambridge Companion to Literature and Environment*, ed. Louise Westling, 169–83. New York: Cambridge University Press.

———. "Source of Life: *Avatar*, Amazonia, and an Ecology of Selves." 2014. In *Material Ecocriticism,* ed. Serenella Iovino and Serpil Opperman, 253–68. Bloomington: University of Indiana Press.

———. 2016. "Humanities." In *Keywords for Environmental Studies*, ed. Joni Adamson, William A. Gleason, and David N. Pellow, 135–39. New York: New York University Press.

Adamson, Joni, Mei Mei Evans, and Rachel Stein, eds. 2002. *The Environmental Justice Reader: Poetics, Politics, and Pedagogy*. Tucson: University of Arizona Press.

Adamson, Joni, and Kimberly Ruffin, eds. 2013. *American Studies, Ecocriticism, and Citizenship: Thinking and Acting in the Local and Global Commons*. New York: Routledge.

Adamson Clarke, Joni. 1992. "Why Bears Are Good to Think and Theory Doesn't Have to Be Another Form of Murder: Transformation and Oral Tradition in Louise Erdrich's *Tracks*." *Studies in American Indian Literatures* 4 (1): 28–48.

Allen, Chadwick. 2012. "A Transnational Native American Studies? Why Not Studies That Are Trans-Indigenous?" *Journal of Transnational American Studies* 4 (1): 1–22.

Appadurai, Arjun. 2001. *Globalization*. London, UK: Duke University Press.
Babcock, Barbara A. 1975. "Why Frogs Are Good to Think and Dirt Is Good to Reflect On." *Soundings* 58 (2): 167–81.
Chagani, Fayaz. 2014. "Critical Political Ecology and the Seductions of Posthumanism." *Journal of Political Ecology* 21: 425–36.
Chakrabarty, Dipesh. 2009. "The Climate of History: Four Theses," *Critical Inquiry* 35 (Winter): 197–222.
Cruikshank, Julie. 2005. *Do Glaciers Listen? Local Knowledge, Colonial Encounters, and Social Imagination*. Seattle: University of Washington Press.
Crutzen, Paul J., and Eugene F. Stoermer. 2000. "The 'Anthropocene.'" *International Geosphere-Biosphere Program Newsletter* 41: 17–18.
Cubitt, Sean. 2014. "Ecomedia Futures." *Journal of Media and Cultural Politics* 10 (2): 163–70.
de la Cadena, Marisol. 2010. "Indigenous Cosmopolitics in the Andes: Conceptual Reflections beyond 'Politics.'" *Cultural Anthropology* 25 (2): 334–70.
DeLoughrey, Elizabeth, and George Handley, eds. 2010. *Postcolonial Ecologies: Literatures of the Environment*. Oxford: Oxford University Press.
Dillon, Grace L., ed. 2012. *Walking the Clouds: An Anthology of Indigenous Science Fiction*. Tucson: University of Arizona Press.
Ecuador Constitution. 2008. *Asamblea Nacional Constituyente* Title VII, Chapter 2, Secs. 1–7, 134–39. https://www.constituteproject.org/constitution/Ecuador_2008.pdf.
"Final Conclusions Working Group 7: Indigenous Peoples." 2010. Working Paper. In *World Peoples Conference on Climate Change and the Rights of Mother Earth* [Indigenous Peoples' Declaration]. April 30. Cochabamba, Bolivia,: UDRME.
Galeano, Juan Carlos. 2009. *Folktales of the Amazon*. Trans. Rebecca Morgan and Kenneth Watson. Westport, CT: Libraries Unlimited.
Goeman, Mishuana. 2013. *Mark My Words: Native Women Mapping Our Nations*. Minneapolis: University of Minnesota Press.
Haraway, Donna. 2015. "Anthropocene, Capitalocene, Plantationocene, Chthulucene: Making Kin." *Environmental Humanities* 6 (August): 159–65.
Hogan, Linda. 1995. *Solar Storms*. New York: Simon & Schuster.
Huggan, Graham, and Helen Tiffin, eds. 2010. *Postcolonial Ecocriticism: Literature, Animals, and Environments*. New York: Routledge.
Hunt, Sarah. 2014. "Ontologies of Indignity: The Politics of Embodying a Concept." *Cultural Geographies* 21 (1): 27–32.
Hurston, Zora Neale. 1937. *Their Eyes Were Watching God*. Philadephia, PA: J. Lippincott Press.
Interstellar. 2014. Directed by Christopher Nolan. Los Angeles, CA: Paramount Pictures.
Iovino, Serenella, and Serpil Opperman, eds. 2014. *Material Ecocriticism*. Bloomington: Indiana University Press.
Kirksey, Eben. 2010. "The Emergence of Multispecies Ethnography." *Cultural Anthropology* 25 (4): 545–76.
Latour, Bruno. 1993. *We Have Never Been Modern*. Cambridge, MA: Harvard University Press. 1993.
———. 2004. "Whose Cosmos, Which Cosmopolitics? Comments on the Peace Terms of Ulrich Beck." *Common Knowledge* 10 (3): 450–62.
Lempert, William. 2014. "Decolonizing Encounters of the Third Kind: Alternative Futuring in Native Science Fiction Film." *Visual Anthropology Review* 30 (2): 164–76.

Loft, Steven, and Kerry Swanson, eds. 2014. *Coded Territories: Tracing Indigenous Pathways in New Media Art*. Calgary, Alberta: University of Calgary Press.

Martian, The. 2015. Directed by Ridley Scott. Los Angeles, CA: Twentieth Century Fox Film Corporation.

Momaday, N. Scott. 1968. *House Made of Dawn*. New York: Harper and Row.

Murphy, Patrick, ed. 2000. *Further Afield in the Study of Nature-Oriented Literature*. Charlottesville: University Press of Virginia.

Nixon, Rob. 2011. *Slow Violence and the Environmentalism of the Poor*. Cambridge, MA: Harvard University Press.

Ortiz, Simon. 1999. *Men on the Moon: Collected Short Stories*. Tucson: University of Arizona Press.

People of a Feather. 2011. Directed by Joel Heath. Montreal, Quebec: Sanikiluaq Running Pictures Ltd. DVD.

Rader, Dean. 2011. *Engaged Resistance: American Indian Art, Literature, and Film from Alcatraz to the NMAI*. Austin: University of Texas Press.

Ray, Sarah Jaquette. 2013. *The Ecological Other: Environmental Exclusion in American Culture*. Tucson: University of Arizona Press.

Rose, Deborah Bird. 2011. *Wild Dog Dreaming: Love and Extinction*. Charlottesville: University of Virginia Press.

———. 2015. "Ethnography." In *Keywords for Environmental Studies*, ed. Joni Adamson, William A. Gleason, and David N. Pellow, 110–12. New York: New York University Press.

Sanchez, Mikeas. 2013. *Mojk'jäya-Mokaya*. Mexico, DF: Pluralia Ediciones.

Shapiro, Steven. 2005. "Cosmopolitics" In *The Pinocchio Theory*. Blog. Accessed May 16, 2016: http://www.shaviro.com/Blog/?p=401.

6th World, The: An Origin Story. 2012. Directed by Nanobah Becker. San Francisco, CA: Independent Television Service, Inc. (ITVS).

"State of the Climate: Global Analysis—Annual 2015." 2016. *National Oceanic and Atmospheric Agency*. January. Accessed May 5, 2016. http://www.ncdc.noaa.gov/sotc/global/201513#gtemp.

Stengers, Isabelle. 2005. "The Cosmopolitical Proposal." In *Making Things Public: Atmospheres of Democracy*, ed. Bruno Latour and Peter Weibel, 994–1004. Cambridge, MA: MIT Press.

Tedlock, Dennis. 1985. Introduction to *Popul Vuh: The Definitive Edition of the Mayan Book of the Dawn of Life and the Glories of Gods and Kings*, trans. Dennis Tedlock. New York: Simon and Schuster.

Thoreau, Henry David. 1864. *The Maine Woods*. Boston, MA: Ticknor and Fields.

Tohe, Laura. 2009. *Enemy Slayer, A Navajo Oratorio*. With Mark Grey, Michael Christie, Scott Hendricks, Deborah O'Grady. © Naxos Music. Naxos 8.559604. Compact disc.

Trees Have a Mother (The): Amazonian Cosmologies, Folktales, and Mystery. 2008. Directed by Valliere Richard Auzenne and Juan Carlos Galeano. Tallahassee: Florida State University Film School. Available at Films on Demand, item number 40566. http://digital.films.com/play/WNHAND.

Universal Declaration on the Rights of Mother Earth and Climate Change (UDRME). 2010. World People's Conference on Climate Change and the Rights of Mother Earth. 22 April. https://pwccc.wordpress.com/programa/.

Viveiros de Castro, Eduardo Batalha. 1998. "Cosmological Deixis and Amerindian Perspectivism." *Journal of the Royal Anthropological Institute* 4 (3): 469–88.

———. 2004. "Exchanging Perspectives: The Transformation of Objects into Subjects in Amerindian Ontologies Castro." *Common Knowledge.* 10 (3): 463–84.

Walls, Laura Dassow. (2009) *The Passage to Cosmos: Alexander von Humboldt and the Shaping of America*, Chicago, IL: University of Chicago Press.

———. 2015. "Cosmos." In *Keywords for Environmental Studies*, eds. Joni Adamson, William A. Gleason, and David N. Pellow, 47–50. New York: New York University Press.

"World Peoples' Conference on Climate Change and the Rights of Mother Earth: Indigenous People's Declaration" [Draft working paper]. 2010. *Global Research.* April 21. Cochabamba, Bolivia, UDRME. http://www.globalresearch.ca/world-peoples-conference-on-climate-change-and-the-rights-of-mother-earth-indigenous-people-s-declaration/18931.

Young, M. Jane. 1987. "'Pity the Indians of Outer Space': Native American Views of the Space Program." *Western Folklore* 46 (4): 269–79.

Zalasiewicz, Jan, Mark Williams, and Colin N. Waters. 2015. "Anthropocene." In *Keywords for Environmental Studies*, ed. Joni Adamson, William A. Gleason, and David N. Pellow, 14–16. New York: New York University Press.

Part I

Resilience

1 Negotiating the Ontological Gap

Place, Performance, and Media Art Practices in Aotearoa/New Zealand

Janine Randerson and Amanda Yates

Whakapapa

Māori *whakapapa* is an oral transmission of genealogical lines of descent in humans, animals, and plants in a densely layered cosmogony. *Whakapapa* offers up a radical expansion of "the human" to include our larger kinship group: the lizards, birds, and ferns; the rivers, mists, and mountains. Māori environmental scientist Mere Roberts and colleagues describe the telling of *whakapapa* as:

> typically beginning with the origin of the universe and the primal parents, then continuing to trace the descent of living and nonliving, material and immaterial phenomena, including humans.
>
> (Roberts et al. 2004, 1)

Whakapapa establishes an ontological framework that involves a paradigm shift of sorts for Western academia, a turning from a primary concern with semiotic subjectivities to a more holistic account of the vast biomass as connected to the human. A cosmogony of the world composed of bacteria, phytoplankton, ants, whales, trees, and humans, as well as intra- and extraplanetary energies, resource systems, and flows. The Māori term *tangata whenua* ("people of the land") describes a cosmogony where humans and all other biotic and abiotic life are the progeny of Papatūānuku, the Earth Mother. To understand plant and animal *whakapapa* requires knowledge not only of plant and animal names but also of their accompanying narratives.[1]

Artist and theorist Dr. Huhana Smith describes *whakapapa* as a meta-account of descent; *whanaungatanga* stands for the reciprocal relations established as part of a highly complex "socio-cultural-ecological system" (2012a, 2). In this framework, all of our actions occur within, or in borders between, socio-cultural-ecological discourses whether we acknowledge it or not, including the domain of contemporary art and design, as we describe below.

This chapter considers the work of three Māori artists who engage *whakapapa* concerns through creative art and design practice as a generative mode of *whanaungatanga*. Robertson's media artwork *Uncle Tasman—The Trembling Current that Scars the Earth* (2008) brings genealogical narratives to bear on a regional ecological issue of air and water quality in the town of Kawerau.

Figure 1.1 Natalie Robertson. Installation detail. Three-screen video installation, *Uncle Tasman—The Trembling Current that Scars the Earth* (2008). In the exhibition, *He Korowai o Wai* Rotorua Museum. 2008.
Source: Courtesy of Natalie Robertson.

Māori artists often attend to situated tribal issues, yet in Robertson's work the global polity is implicated in the transnational flows surrounding the market for the Tasman corporation's paper products and the complex chain of environmental effects to which their Kawerau factory contributes. Shannon Te Ao's performance artwork, *Untitled (after Rākaihautū)* (2012a) follows the ancestral pathways of the chief Rākaihautū as he molds landscapes and lakes with his footprints, according to oral tradition.

Figure 1.2 Video still from Shannon Te Ao, *Untitled (after Rākaihautū)* (2012a). The performer Te Ao in Waimea estuary, Whakatū (Nelson) at the top of New Zealand's South Island. Te Ao is filmed by cinematographer Iain Frengley.
Source: Courtesy of Shannon Te Ao.

By mark-making with digging sticks in the Nelson mudflats at the top of New Zealand's South Island, Te Ao draws attention to the seabed and foreshore as a site of ongoing contention over sovereignty rights between the government and the Māori. Amanda Yates's *Pop-up Garden / Māra Iti* (2012) aims to reshape urban landscapes by bringing edible gardens into the city.

Working with the Wellington City Council and referencing Māori urban forms that bind landscape and architecture together, Yates utilizes the pop-up garden as an eco-political tool and change agent. We will argue that the respective practices of Robertson, Te Ao, and Yates build relationships (*whakawhanaungatanga*) between entities and entity-environments through sensitive, forceful, and questioning engagements in the world.

We propose that the video works, performances, and landscape actions are cultural expressions that attempt to bring ontological difference to the fore. The "ontological turn" in recent humanities and social science discourse has been questioned by some Indigenous researchers (Hunt 2014; Todd 2014), yet we argue that this "turn" as articulated in contemporary

Figure 1.3 Amanda Yates. *Pop-up Garden / Māra Iti* (2012), Wellington. Source: Courtesy of Amanda Yates.

philosophies can be supported by the sustained position of the nonhuman in Indigenous philosophies. We begin our discussion by clarifying the debate and our proposal, which builds on Hunt's notion of the productive ontological "gap." Then, turning to the work of these Māori artists, we suggest that art and design projects can activate change in the audience through discussion, sharing food, or affective response. Sensorial approaches to nonhuman entities such as mud, mist, steam, or plants probe "gaps" (Hunt 2014) within normative cultural frameworks where socio-ecological change can be imagined.

Ontology and Indigeneity

To operate within the frame of *whakapapa* requires an ongoing questioning of how Indigenous concepts are being discussed, and for whom, to ensure that such accounts remain mobile and relational. We will preface our account of specific cultural/creative practices by examining both ontology and Indigeneity in the context of the ontological turn across the social sciences and humanities. Movements from various disciplines have foregrounded the agency of things or assemblages of human and nonhuman and a turn toward material and multispecies relations (Adamson 2012; Barad 2007; Bennett 2010; Braidotti 2013; de la Cadena 2010; Haraway 2008; Latour 2004a; Stengers 2005; Whatmore 2002). These authors share concerns around the nature of being and provoke a radical reshaping of how various fields have conceived of human and nonhuman agency, change, materiality, and cosmopolitical, ecological, multispecies, and object relations. Multiple worldings, rather than "*the one* cosmos" or a mononaturalist approach (Latour, 2004b, 453), meet in the practices we will discuss.

Current ontological concerns parallel Indigenous thought around nonhuman agency and material vitality, embraced in the aforementioned concepts of *whakapapa* and *whanaungatanga*. Anthropologist Marisol de la Cadena (2010), while writing of Peruvian Indigenous practices, draws on philosopher of science Isabelle Stengers' (2005) cosmopolitical stance that advocates a "slowed-down reasoning" in relation to the sacred mountains, known by name as earth-beings. Like Rūaumoko, the god of earthquakes in Māori cosmology, who is referenced in Natalie Robertson's work (2008), the Andean mountain-being Ausangate demands respect, or he may become dangerous (de la Cadena 2010). This lack of separation between nature and humanity is contentious when incompatible ontologies, such as corporate forestry interests and Indigenous cosmogonies, clash over issues such as air and water rights. For instance, the representation of an alternate Māori cosmos in New Zealand director Niki Caro's film *Whale Rider* (2002) foregrounds a multispecies relation (Adamson 2012, 35). The Ngāti Konohi *iwi* regard the whale-shaped island in Whangara as part spirit (*tīpua*), part human (*tangata*), and part *taniwha* (whale or water-being)—qualities also transmuted into geological form. Yet Robertson also comments (2011) that, tragically, the island, Te Ana-o-Paikea, had its tail cut off and removed for road gravel.

Some Indigenous scholars, working with similar concepts to vitalism and agential matter within Indigenous knowledge systems, have observed the ongoing "turn" occurring in sections of EuroAmerican academia with interest and some irony. Social anthropologist Zoe Todd, an Indigenous Canadian, in critiquing the (neo)colonial time-lag for non-Indigenous academics, tells us to think deeply about how:

> the Ontological Turn—with its breathless "realizations" that animals, the climate, water, "atmospheres," and nonhuman presences like ancestors and spirits are *sentient* and *possess agency*, that "nature" and "culture", "human" and "animal" may not be so separate after all—is itself perpetuating the exploitation of Indigenous peoples.
>
> (Todd 2014)

In addition, Indigenous researcher Sarah Hunt (2014) argues that "the potential for Indigenous ontologies to unsettle dominant ontologies can be easily neutralized as a triviality, a case study, or a trinket, as powerful institutions work as self-legitimating systems that uphold broader dynamics of (neo)colonial power" (30). We share this wariness as writers of both Māori and Pākehā (European settler) descent, but also share a sense of optimism about the possibility of redress of historical injustice that an alignment between Indigenous thought and the ontological turn might produce. The creative strategies discussed below operate from an implicit concern to counter colonial accounts of place and history.

The Negotiated Gap

The value of visual, aural, and performative storytelling as history, representation, and genealogical account is part of our reading of the creative works. The stories occupy multiple temporalities and begin to heal the violence of boundary practices that are asserted through the "mononaturalist" representations of landscape and "land as resource" (Watson 2014, 76). Hunt (2014) highlights a gap between Indigenous and modern ways of knowing while identifying the porous, elusive nature of knowledge itself. Describing an engagement with the ritual practices of the *Kwakwaka'wakw* potlatch, Hunt notes that:

> the ontological differences are difficult to explain, yet that is where their power lies—in the spaces between intellectual and lived expressions of Indigeneity. I would propose that these gaps in regimes of knowledge provide sites where ontological shifts are possible. So how do we better expose and explore these gaps? [...] One starting place might be accepting the partiality of knowledge. Its relational,

> alive, emergent nature means that as we come to know something, as we attempt to fix its meaning, we are always at risk of just missing something (2014, 30–31).

The gap, for the purposes of our argument, is the space in-between, a space of negotiation that exceeds human language and breaks away from a hermetic concern with human behaviors toward other ways of knowing. We suggest that the material and performative practices of art can operate in an unfixed gap, although art is never an unmarked, apolitical space. Many art museums locally and internationally have exhibited the moving image work of Robertson and Te Ao, and Yates's *Pop-up City / Māra Iti* was sited in a public square in partnership with the Wellington Council in our capital city; so there is a relationship between artists and mainstream institutions in postcolonial Aotearoa/New Zealand. Yet creative activity occupies a differentiated zone where political resistance can be expressed in free play. Whereas informed Māori protest marches known as *hīkoi* (Nathan 2015) are often portrayed by the mainstream media as the radical actions of a vocal few (New Zealand Herald 2011), art is often accepted as less threatening than street protest, but the implications are no less radical. By drawing pre-colonial customs into contemporary reshapings or representations of the land or landscape, Robertson, Te Ao, and Yates reveal the partiality of bicultural exchange. The Māori cosmopolitical position that things have *mauri* ("life force") is a useful prompt to question the perceived fissure in non-Indigenous ontologies between subject and object, human and the nonhuman.

Natalie Robertson, *Uncle Tasman—The Trembling Current That Scars the Earth*

While the academic tradition has historically presented itself as having "no geopolitical determinations" (as noted by Gayatri Spivak 1988, 24), Indigenous academics often speak from a grounded ethics of care both for local polities and the wider implications of care for climate or environment. Over the last decade Ngāti Porou/Clann Dhonnchaidh artist Natalie Robertson's practice has exemplified Smith's (2012b) assertion that the restoration of fragmented ecological systems is interdependent on the healing of both human and nonhuman communities. Robertson uses video and sound as a means of bringing *whakapapa*, genealogical narratives and care for the ecosystem (*kaitiakitanga*), to bear on the economic imperatives of the pulp and paper industry. She interrogates the gap created by the colonial or scientific regard of mountains or rivers as "natural resources" that are out-of-sync with Māori relations to ancestor-beings.

Robertson's current work focuses on the Waiapu River, a tribal polity on the east coast of Aotearoa/New Zealand's North Island, a waterway near her east coast *hapū*, which has been degraded as a result of heavy deforestation.

However, the three-channel video art installation *Uncle Tasman—The Trembling Current That Scars the Earth* (2008–2012) is set in the region of Kawerau, Robertson's inland birthplace. The camera frame is steadfastly fixed on an ancestor mountain (Pūtauaki, adjacent to Kawerau), a waterfall (Tarawera) and a fumarole at Whakaari (an active island volcano in the Bay of Plenty).

Figure 1.4 Natalie Robertson. Still image of Pūtauaki mountain, from the video installation *Uncle Tasman—The Trembling Current that Scars the Earth* (2008). Source: Courtesy of Natalie Robertson.

Figure 1.5 Natalie Robertson. Installation detail of Pūtauaki mountain and Whakaari fumarole. *Uncle Tasman—The Trembling Current that Scars the Earth* (2008), Toi Rerehiko Moving Image Centre, Auckland. Source: Courtesy of Natalie Robertson.

The changing cloud and steam conditions of these sites veil and unveil them, suggesting that they are watched, protected, under the patient gaze of a *kaitiaki* (spiritual guardian or caretaker). These three places retain their *mauri* and contain physical reminders, or memories of an adjacent geothermal Lake Rotoitipaku before it became heavily polluted by the activities of the Tasman Pulp and Paper Mill (nicknamed "Uncle Tasman" by local residents). In Ngāti Tuwharetoa's accounts, Lake Rotoitipaku, the absent subject of the artwork, is the birthplace of the warrior Tuwharetoa and the burial place of his mother, an important descendent from the Arawa waka (that traveled from Hawaiiki). The spring that welled up in the lake, known as "the bubble," is a sacred spring. The lake was well loved by Kawerau inhabitants and was a safe place to swim once the Tarawera River was polluted by the mill. But the calamitous dumping of waste by the Tasman mill from 1968 onwards ended the health of the body of water. The artwork is a memorial to Lake Rotoitipaku. Robertson now describes the sacred place as a "sludge pond" (2011); it serves as a visual gap in the spectacle of her film, although it is clearly present in the audio composition.

Robertson's work operates in the partial, or dimly understood, space between the Western form of "landscape" photography and the dense narrative layers of a Ngāti Tuwharetoa ontological frame in the vocal and environmental soundtrack of the work. The audio track begins with the constant sound of the waterfall in the Tarawera River, coupled with a *tangi* lament performed by a woman. The waterfall of the Tarawera River in *Uncle Tasman*, represents the tears of Tarawera as she mourns "for the destruction of her *mauri*" (Robertson 2011). The vocal layer in *Uncle Tasman* later commences with a *tauparapara*, a chant by Te Haukakawa Te Rire that traces the Ngāti Tuwharetoa of Kawerau's tribal histories of seismic and geothermal activity that describes the love affairs of the three volcanic cones that frame the region. The second vocal element is Wayne Huia Peters' *korero* ("dialogue") in English. He tells his story from the personal perspective of being a young child who would swim and fish in the "food basket" of Lake Rotoitipaku, which was a haven compared to the polluted Tarawera River. But after the dumping of waste by "Uncle Tasman" in 1968, the fish and eels had nowhere to escape, eventually crawling onto the road to die. He remembers this sight as "soul-destroying," as he witnessed the *mauri* of the lake being devastated (Robertson 2011). The Kawerau tribal elders provided Robertson with access to record these local stories, some of which were part of the evidence brief read to support Waitangi claims brought before the Waitangi Tribunal, a permanent commission of inquiry charged with making recommendations on claims brought by Māori relating to actions or omissions of the Crown (New Zealand Government) that potentially breach the promises made in the Treaty of Waitangi (1975) (see New Zealand Ministry of Justice, n.d.).

Towards the end of the work, the aural narrative of *Uncle Tasman* moves through grief and loss to the assertion of ecological restoration in the rousing sound of the *haka*. This is the customary Ngāti Porou *haka*, performed by women, to establish Robertson's right to tell the Ngāti Tūwharetoa story of Lake Rotoitipaku and to demand social and environmental restitution. Robertson explains the purpose of the *haka* as follows:

> The customary Ngati Porou haka Ka Panapana is performed to call for social justice and is followed by a haka. Rūaumoko, who is God of earthquakes and volcanoes, and Papatūānuku's (Earth Mother) unborn child, is stirring and kicking, eager to live up to the promise of its name, signifying an earthquake scar, a reminder that Aotearoa sits balanced along the Pacific "rim of fire" with potential for sludge from deeply toxic dumping sites to be scattered far and wide from what is known as the Taupō volcanic plateau if eruptions were to occur from that site.
>
> (Robertson 2007)

A human presence is maintained throughout the work in the audio track, yet we never see an image of a person, only the locked-off shots of the three enduring land features that are often dissolved by cloud. In *Uncle Tasman,* the cloud is from four sources; steam from the boreholes and hissing fumaroles of Papatūānuku, the atmospheric mist and cloud of Tāwhirimātea (the weather god), the steam rising from the geothermal lake at the foot of Pūtauaki, and the toxic, human-generated steam and smoke from the chimney stacks of the mill. The cloud obscures and then clears to reveal the landforms and factory in the foreground of the center frame, much like the hidden story Robertson conveys. Cloud or mist is a transitive element indicating movement, change, and resilience, as well as a formal device for Robertson to counter the stillness of the photograph, yet retain the control of a locked-off camera position (Robertson 2011). Robertson's clouds have a strategic function in the pictorial order as "an integrator and a disintegrator, now as a sign, now as a non-sign" (Damisch 2002, 184). Clouds operate in the intermediary, obscured patch in vision where solid forms become unfixed and alive. Isabelle Stengers (2011) suggests traditional (nonmodern) cultures are often unfairly associated with nonnegotiable prohibitions (2011, 369). Yet the active nature of cloud in Robertson's video suggests Indigenous values that are responsive to change, as opposed to stereotypes, about "nonmodern culture" that suggest that they are static or inflexible.

The video installation has international resonance, as its wide exhibition in Shanghai, Paris, and San Jose (CA) attests, yet the most politically sensitive exhibition of the work from Robertson's perspective was at the Whakatāne District Museum, the nearest gallery to Kawerau in Aotearoa/New Zealand. The local newspaper, the *Whakatāne Beacon,* reported on the exhibition under the headline "Uncle Tasman's Abuses Investigated through Art" (Miller 2008), drawing attention to the highly politicized context of the artwork.

Much of the local Māori community is employed by the Tasman mill, which in 2009 was granted permission by Environment Bay of Plenty to continue discharging effluent into the river, as well as gas and dust emissions. This permission extended a right—first granted in a 1954 Act of Parliament—for another twenty-five years (Bay of Plenty Regional Council 2004; Davison 2009). Community development councilor Materoa Dodd (2010) argues that the pollution of the Tarawera River, now often referred to as "the black drain" has effected three *iwi* ("tribes"): Te Arawa, Tuwharetoa ki Kawerau, and Ngāti Awa. The destructive effects of forestry include the acidification of the soil, the leaching of organic chlorides into rivers, and the transmission of dangerous airborne dioxins to the mill workers and local inhabitants. Furthermore, the Tasman mill has been linked to cancer-causing dioxin poisoning for mill workers and the town's inhabitants, the majority of whom are Māori. Kawerau now has recorded the highest age-related cancer registration rate of all districts in New Zealand. (Bay of Plenty District Health Board 2005, 28) For Dodd, the lack of regard by the Tasman Pulp and Paper Company has "forced a fundamental change on the *iwi*, a change that has affected their cultural and social relationship to the river. The price of economic development on the river has been cultural and ecological genocide." (Dodd 2010).

Yet, in respecting the complex relationship its local audience has with the mill, *Uncle Tasman* does not explicitly link the airborne dioxins and cancer deaths to images of polluted rivers. Instead, the focus on the oral account of Lake Rotoitipaku's demise presents a "tangible and unarguable" event by which to politicize the ecological ills perpetrated by the Tasman Company (Robertson 2011). Inviting its Māori community as well as those beyond it to engage with it, this artwork prompts into existence new and formerly unacknowledged dimensions of a complex event. Robertson's video might be considered a place of ontological encounter, a mode of operation where alterior or in-between positions can be acknowledged and discussed beyond the conventions of mainstream media or direct political activism. Another Māori artist who similarly unfixes the frame of "landscape" and invites the concept of a negotiable gap is Shannon Te Ao.

Shannon Te Ao: *Untitled (after Rākaihautū)* (2012a)

A man clad in black jeans slides across a muddy foreshore on an overcast day. He leans down and sinks his hands into the mud. At first he tentatively pats the shallow water of an outgoing tide. Then he begins to crawl across the mud on all fours, slapping it loudly. He takes a stick and begins to scrape marks forcefully in the mud. The man uses his hands to scrape the mud and forcefully molds the mud into mounds. Finally, he takes a stone and throws it at the mud in apparent frustration. Gradually his frenetic movements become gentler; he crawls again on his knees, smoothing the mud and water. He digs a channel through the mud. He molds a long hillside. In the final wide-angle shot, his reflection is mirrored in the shallow water of the tide as he walks away.

Figure 1.6 Video still from *Untitled (after Rākaihautū)* (2012a). Te Ao in Waimea estuary, Whakatū (Nelson) at the top of New Zealand's South Island. Te Ao is filmed by cinematographer Iain Frengley. Source: Courtesy of Shannon Te Ao.

The figure is artist Te Ao digging in Waimea estuary, Whakatū (Nelson), at the top of New Zealand's South Island. Te Ao is filmed by cinematographer Iain Frengley.The Waimea inlet is a large open space beside the fast-growing urban centers of Nelson and Richmond. The estuary is a dominant landscape feature at the head of Tasman Bay. In this work Te Ao examines the estuary, often regarded as a neglected no-man's land or (we suggest) as an ontological gap, where a shifting place-identity can be enacted. This gap relates to the literal site of the estuary as in-between the sea and the land but also to the sociopolitical and cultural gap between colonial New Zealand and *mātauranga* Māori connections to land. Te Ao's foreshore becomes a dialogical, negotiable space by activating the wet, muddy ground, transforming it from inert wasteland (as it has been perceived in the industrial age) to living being. The foreshore as an unfixed space is an appropriately permissive site for performance art, where histories can be redressed and problematized. *Untitled (after Rākaihautū)* (2012a) was a pivotal piece in Te Ao's emerging practice and has been widely exhibited nationally in New Zealand (Te Tuhi, Auckland, Te Papa, Wellington; the Physics Room, Christchurch) and internationally in Taipei.

Robertson's practice aimed "to remember, name, witness, and identify" (Robertson 2007) by shifting between historical tribal and contemporary narratives. Similarly Te Ao's video work is bound to the narrative of a famed, historical seafarer, Rākaihautū, and the mud he digs becomes an active agent in an eco-political debate that calls up this story. Te Ao himself is featured in the video, immersed in a figurative activation of this cultural story, unlike Robertson, who presents a slow meditation on the ancestor-beings.

Rākaihautū was the first explorer of Te Wai Pounamu (the South Island of New Zealand), and as recounted by Te Ao, he circumnavigated the island

and beat the earth with his *kō* ("digging stick"). The Nelson Lakes are said to have formed as the result of these actions. In 1918, H. Beattie (1918, 236) recorded the following history of Rākaihautū from the Waitara *iwi*:

> Ancient story. Now! Let all people of every part know, Rākaihautū was the man and "Uruao" the canoe [...] His was the first canoe to come to this island. [...] The proverb *Ka puna karikari a Rākaihautū* ("The springs dug out by Rākaihautū") refers to the seas (lakes) lying in this island, and the saying, *Ka poupou a Te Rakihouia* ("The posts of Te Rakihouia"), refers to the *pa-tuna* ("eel-weirs") in the rivers on the eastern and southern sides of the island. Te Rakihouia came along the seashore and erected the posts. [...]

The story of Rākaihautū's landing in Aotearoa parallels the artist's own recent migration. Te Ao grew up in Sydney, Australia, then voyaged back to Auckland, Aotearoa's largest city, and at the time of making this work, he had just moved to Nelson in the South Island. When Te Ao was welcomed at his local *marae,* he made the spatial transition from *waewae tapu* ("a newcomer") to *tangata whenua* ("people of the land") by crossing from one side of the *marae* to another. The element of the temporal continuity between present day and ancestral journey frames the event of the work (Te Ao 2012b).

Te Ao selected the location for the performance as the closest position to that of Rākaihautū's actual landing. Yet the narrative of a significant ancestor is transformed as Te Ao reframes the arrival of Rākaihautū as an ongoing performance rather than a defined moment in time. In allowing for this continuity, the artwork attends to the gap between generations through a cyclical temporality. Anthropologist Marilyn Strathern, an early harbinger of the "ontological turn," writes the following in relation to the Western view of linear, historical time and a Melanesian concept:

> An event [happening] taken as a *performance* is to be known by its effect: it is understood in terms of what it contains, the forms that it reveals or conceals [...] Consequently, time is not a line between happenings; it lies in the capacity of an image to evoke past and future simultaneously.
>
> (Strathern 1990, 26–29)

The video/performance materializes the circularity between past and present as Te Ao looks forward to his arrival in Nelson by recalling the past. His mark-making re-inscribes the history of the Waimea estuary with the wounds of the contemporary moment.

These wounds point to the ecological deterioration of the Waimea estuary and the associated cultural politics of the site. According to the council management plans, significant loss of stream-and-shore marginal wetland and saltmarsh has reduced the natural filtering of the ground, and parts of the inlet are probably muddier than in earlier times. The Waimea inlet is of international importance for migratory bird species and is of national significance for other endangered or threatened species. Storm water outlets from urban areas are a major source of contaminants causing the shellfish, once a major source of food for the Indigenous occupants, to become unfit for consumption (Nelson City Council 2010). Shannon Te Ao states:

> The site and the performance both address models of understanding a place through our physical relationship to the land. This would seem to be fundamental in relation to discussion around the Foreshore and Seabed debate. The nature of the tidal estuary itself is of a transitional, temporal space. Metaphorically, ideal for ongoing dialogue.
>
> (Te Ao 2012a)

The Foreshore and Seabed debate, alluded to in Te Ao's statement above, is still a site of contention in bicultural politics in Aotearoa/New Zealand. In 2003 the New Zealand Court of Appeal granted a significant decision on Māori land title in favor of a *hapū* in relation to customary ownership of the foreshore and seabed. This resulted in a "tsunami" (Ruru 2004) of public outrage, where the New Zealand white majority felt that this meant that the beach and coastal regions would now be policed by Māori interests and the foreshore made inaccessible. The government, in contradiction of the courts, announced that they would legislate that the seabed and foreshore was Crown land exclusively.

Māori, in a stand of unity, rejected the proposed legislation that was rushed through under urgency, and enacted a *hīkoi* ("protest march") to the New Zealand parliament in Wellington. As Māori legal commentator Jacinta Ruru argues, in a Māori worldview there is no distinction made between the land underwater or seabed or dry land; it is all regarded as part of a garden or food basket. The intertidal zone was particularly significant for shellfish gathering as a major source of food (Ruru 2004, 58). Several local *hapū* sought a status order of land ownership, and the legislation was subsequently amended, but the debate is regarded as typical of Pākehā alienation of Māori land.

Through his immersion in the material of mud itself, Te Ao seems to express the identity of *tangata whenua;* through *haptic* actions, he is of the land. The muddy scene could be read as abject without the cultural framework. Mud, often associated with the unclean in Western culture, in this context suggests a healing ritual that connects past and present. The tele-documentary tradition of a clean historical re-enactment of a cultural story is subverted and radicalized. Te Ao's actions retain a sense of disturbance, trauma, and the violence of migration and colonial disenfranchisement. Several authors on the work have commented on the sense of frustration (Smith 2012b; Te Ao 2012a) apparent in the striking of the earth. Smith writes:

> A slightly harrowing experience for the artist, we follow his tracing of footfall over an estuary and witness his slapping, scraping, stabbing

> with a stick, raising and shaping of gravels, flicking of mud and *haehae*-like scarifications that he leaves in Waimea estuary, Whakatū (Nelson), when he walks out of frame.
>
> (Smith 2012b, n.p.)

This apparent aggression might seem at odds with the values of *kaitiakitanga*, ("conscience"), and symbiotic relationships with the earth that are increasingly "expected" of Māori by other New Zealanders. We could read the violent actions as a rejection of the mantel bestowed on Māori to be responsible for the land's health, where successive predominantly Pākehā governments are failing. According to these writers, the responsibility of care for our foreshore should be the collective responsibility of Māori and Pākehā New Zealanders.

Negotiating these collective responsibilities through the invitational gaps of Te Ao and Robertson's artwork might open up painful histories for artists and their audience, but notions of environmental healing are also central. Next we turn to Amanda Yates' project *Pop-up Garden—Māra Iti*, which operates in a playful space to welcome such negotiations of collective responsibility.

Amanda Yates's Pop-Up Garden—*Māra Iti* (2012)

While Robertson and Te Ao's projects attend to the environs of small towns, Amanda Yates' installation *Pop-up Garden—Māra Iti* extends an ethic of care to urban dwellers, in the form of freely available sun-ripened fruit, salad greens and herbs in Aotearoa/New Zealand's capital city, Wellington. Developed with the support of the Wellington City Council the project inhabited the Civic Square with large-scale planters filled with heritage tomatoes, scarlet runner beans, rainbow chard, peas and herbs: the *Māra Iti* installation aimed to act as "eco-acupuncture," a small scale intervention enacting a more healthful, resilient and locally-referenced urbanism. Yates organized public gardening workshops for adults and children to grow conversations around resilience and urban community building.

Wellington's urban form is concentrated, held between sea and hills, the city's strategic plan emphasizes a people-centered, resilient eco-city as a central vision. The garden emerged from an urban research lab that Yates ran for the Wellington Council's Urban Design unit, and often with Massey University Spatial Design students: in these projects the city was engaged as a living laboratory within which to question, speculate and, in the case of Yates's practice, test out social and design innovations. Working with student practitioners, local activists, neighbours, Council urban and landscape designers, parks people, gardeners, and engineers refracted the diversity of the city, creating a variegated network that explored issues of resilience and sustainability.

As a collaborative work, *Māra Iti* embraced an Indigenous mode of hospitality, *manaakitanga*, offering an ethic of care to those who use the square. Six 2.2 meter high frame structures were placed along the edge of the Civic Square's raised lawn, where local white-collar workers often sit to eat their lunch.

Figure 1.7 Amanda Yates. *Pop-up Garden / Māra Iti* (2012), Wellington. Source: Courtesy of Amanda Yates.

As soon as the planters were installed people grouped around them to find and eat sun-warmed peas straight from the pod. The planters lensed how urbanism requires a civic contract, where strangers are considerate of others, largely managing their behavior in response to what the culture expects of them. This collective consideration manifested in subtle ways, with people harvesting selectively, rather than taking everything; one man returned over the months of the installation, carefully harvesting the rainbow chard, taking only a few leaves of an adequate size. A local resident watered and tended to the planters, acting as an informal host to those interested in the gardens. This careful guardianship of a group resource demonstrates a contemporary form of *kaitiakitanga*.

In addition, over the summer the plants quickly started to supersede the confines of the architectural frame, climbing up string structures, and turning the planters into a green tower, while further vegetal shoots stretched out along the ground, extending the periphery of the planters outwards. Impelled by a narrative of massive anthropogenic ecological change *Māra Iti* enacted at the small scale an urban-rural continuum influenced by a Māori cosmogony of connection, where humans descend from primordial parents of Ranginui (sky) and Papatūānuku (earth), and are related to the material and immaterial world. Māra Iti sits within Yates' larger project to explore an ecological urbanism that pays particular attention to located and cultural forms of knowledge: this "Indigenous" urbanism addresses resource generation, healthful and child-oriented design, and blue-green "infrastructures" within a Māori socio-cultural-ecological (Smith 2012a) framework.

Māra Iti as an ecological design articulates a resilient and generative urbanism by literally returning whenua (land) to productive use-value for *kai* ("food"). Like Robertson and Te Ao's work, the garden project also emphasizes a recursive ontology by returning to the ancient use of whenua in a public site adjacent to where food was once grown by Māori occupants. Such projects take steps towards restoring the socio-cultural-ecological health of fragmented Māori and non-Māori urban communities.

Conclusion

Māori concepts such as *kaitiakitanga* ("guardianship") suggest a sensitive attending to *whakapapa* ("the relations with ancestor-beings") and offer a rich account of ecological relationships and reciprocity that can be brought to bear in media artworks or installation practice. This Māori worldview, increasingly shared by Pākehā New Zealanders, acknowledges that humans are only one element of the given world. In 1991 the New Zealand government's Resource Management Act recognized *kaitiakitanga* in a limited way, as a concept for the co-custodianship of physical resources between iwi and local authorities. For many Māori, *kaitiakitanga* is the obligation to their *tīpuna* ("children") and to their *mokopuna* ("grandchildren") (Kanewa 2010, 1). The cultural paradigm of *kaitiakitanga* implicitly critiques anthropocentric utilitarianism. The short-term horizon of capitalism is temporally out of step with the Māori conception of an obligation to the future that is part of *whakapapa,* or multispecies genealogy.

At a point in Aotearoa/New Zealand's history where the cataclysmic effect of European colonization and the systematic exploitation of local ecologies and Indigenous culture is only just beginning to be acknowledged by Pākehā in academia (Park 2007) and in government organizations (Kura Kaitiaki Water Conservation project), the promotion and revival of *kaitiakitanga* ("guardianship") is crucial. Artists such as Robertson herself became a *kaitiaki* ("caretaker") of sorts for her tribal polity in Kawerau when her tribal elders were urged by Annette Sykes, a prominent Māori lawyer and Ngāti Porou activist, to tell their stories of the degradation of Lake Rotoitipaku to her. In an interview, Robertson describes her artworks as concerning "relationships between people, the land, and the atmosphere, not simply about human politics *or* ecological issues" (Robertson 2011). To create the work required a long social process where Robertson engaged with and recorded kaumātua and *kuia* (the elders), environmental activists and other town inhabitants. This is a process referred to earlier in the chapter as *whakawhanaungatanga*, or acting with care by building relationships between entity-environments and people. As Smith (2012a) writes, "*kaitiaki* had to actively address the disjunctures among genealogically related people," such as easing difficulties between *whānau* and *hapū* members over land tenure (24).

Although many New Zealanders embrace concepts such as *kaitiakitanga* as a bicultural reality, ecologists and Māori stakeholders often critique the motives of the New Zealand government, which has a history of conservationist protectionism (Park 2007). The parceling of land as sanctuary can lock out customary practices and disallow the expression of spiritual relationships with the nonhuman. Simultaneously, the colonial project and present industries continue to exploit Aotearoa/New Zealand's remaining environmental resources and the boundaries of National or Forest parks are reframed at the will of governments and industry (Davison 2014). Ecologically-inflected Māori art and design projects can maintain these issues in the public ear and eye as a form of extended *kaitiakitanga*.

While Robertson, Te Ao, and Yates occupy the role of *kaitiaki* ("guardians") differently, they all address human relationships and the particularities of the sites they occupy. The assumption of an "inherent" Māori environmentalism can be detrimental, as literary critic Laura Wright argues, "if one considers the colonial rhetoric of the noble savage as a being somehow at once pure and uncivilized, more connected to nature" (Wright 2010, 74). Rather, we are exploring a place-based ontology of connection that calls for a continually enacted ethic of care across settler and Indigenous cultures in Aotearoa/New Zealand. As Smith suggests:

> If pollution, contamination, or desecration affected the *mauri,* or life vitality, of revered places within the natural environment, then this influenced the communities who relied on the land, often manifesting in disquiet, disunity, or fragmentation within those communities.
>
> (Smith 2012a, 16)

The decline in the *mauri,* or life force health of the land, as a source of cultural dysfunction is perceptible in Te Ao's striking actions in the estuary, it is directly expressed in the stories and chants in Robertson's work and is a stimulus for Yates's *Mari Iti* garden installation in the urban square.

These Māori artworks return to *whakapapa* as a cosmogony of connection in order to question the normative, modernist frameworks where humans are viewed as somehow outside or beyond the physical world. Art practice can experiment with alternative systems or modes of representation of critical issues such as climate change, resource depletion and land and water pollution that arise from industrialization and state tolerance of environmentally detrimental practices. In the "age of the Anthropocene" (Braidotti 2013, 5)—proposed as the moment where the human has become a geological force capable of affecting all life on this planet—and the looming possibility of the post-Anthropocene (Bratton 2013), the artists discussed here leverage their work as cosmopolitical and potentially transformative tools. Stengers' version of cosmopolitics is "neither an activity, nor a negotiation nor a practice, but the mode in which the problematic co-presence of practices may be actualized" (2011, 371–72). The often problematic co-presence of colonial and Indigenous practices of ecological belief are negotiated through the art and design works we have discussed.

Sarah Hunt's notion of the "ontological gap" is the zone of encounter where the "gap" is a creative, negotiable space where worldings can meet. Moments of ontological slippage operate between mainstream and Indigenous cosmogonies, where performative modes of creative practice offer alterior experiences of familiar places and the common framing of environmental issues. We have suggested that the temporal and perceptual shifts, generated through moving image and performative actions, can engage audiences in unanticipated ways. Sensorial reorientations occur through intensive scrutiny of *te mea*—"the things," or stuff of place—Robertson's long regard of a mountain, Te Ao's bodily immersion in mud, or food that shares itself by growing in an urban environment.

Glossary

kaitiakitanga (noun): guardianship, stewardship
kumara (noun): sweet potato, a root vegetable brought from Polynesia to Aotearoa/New Zealand
mauri (noun): life principle, vital essence, special nature
whakapapa (noun): genealogy, genealogical table, lineage, descent
whakawhanaungatanga (noun): process of relationality, of engaging considerately with others

Note

1. Following these natureculture understandings, we use the words "Indigeneity" and "Māori," as well as "environment," within the chapter as mobile things, avoiding the fixing of these terms as if they were stable and whole categories. Robertson (2011) emphasizes that "Māori" is only one term for the Indigenous New Zealanders. Cosmological accounts of being differ even between *iwi* ("tribe") and *hapu* ("extended family group"). In reference to the relatively recent term "Indigeneity," Indigenous Canadian researcher Sarah Hunt suggests that Indigeneity might be understood as coming into being through "heterogenous, place-based ways of knowing" (2014, 29) for which we will argue that *whakapapa* is one mode of expression.

References

Adamson, Joni. 2012. "Whale as Cosmos: Multispecies Ethnography and Contemporary Indigenous Cosmopolitics." *Revista Canaria de Estudios Ingleses* 64 (1): 29–45.

Barad, Karen. 2007. *Meeting the Universe Halfway: Quantum Physics and the Entanglement of Matter and Meaning*. Durham, NC: Duke University Press.

Bay of Plenty Regional Council. 2004. *Tarawera River Catchment Plan: Summary of Issues*. 1 February. Accessed December 20, 2016. http://www.boprc.govt.nz/media/30995/Plan-040201RegionalPlanForTheTaraweraRiverCatchmentChapter_9.pdf.

Bay of Plenty District Health Board. 2005. *Bay of Plenty District Board Health Needs Assessment: Health Status Analysis 2001–2005*. September 2005. http://s3.amazonaws.com/zanran_storage/www.bopdhb.govt.nz/ContentPages/2324627753.pdf.

Beattie, H. 1918. "Traditions and Legends. Collected from the Natives of Murihiku. (Southland, New Zealand)." *Journal of the Polynesian Society* 27 (107): 137–61.

Bennett, Jane. 2010. *Vibrant Matter: A Political Ecology of Things*. Durham, NC: Duke University Press.

Braidotti, Rosi. 2013. *The Posthuman*. Malden, MA: Polity Press.

Bratton Benjamin. 2013. "Some Trace Effects of the Post-Anthropocene: On Accelerationist Geopolitical Aesthetics." *e-flux*. Journal #46. http://www.e-flux.com/journal.

Cheng, Derek. 2011. "Hīkoi Hits Out at Māori Party." *newzealandherald.co.nz*. March 23. http://www.nzherald.co.nz/nz/news/article.cfm?c_id=1&objectid=10714312.

Damisch, Hubert. 2002. *A Theory of /Cloud/: Toward a History of Painting*. Trans. Janet Lloyd. Stanford, CA: Stanford University Press.

Davison, Isaac. 2009. "Mill Gets 25-Year Pollution Consent." *newzealandherald.co.nz*. October 16. http://www.nzherald.co.nz/business/news/article.cfm?c_id=3&objectid=10603488.

———. 2014. "PM Defends New Gas and Oil Exploration Permits." *newzealandherald.co.nz*. April 9. http://www.nzherald.co.nz/business/news/article.cfm?c_id=3&objectid=11235217.

Dean, Nathan. 2015. "Hikoi to Protect NZ's Seabeds Reaches Waitangi." *Māori Television*. 6 February. http://www.maoritelevision.com/news/regional/hikoi-protect-nzs-seabeds-reaches-waitangi.

de la Cadena, Marisol. 2010. "Indigenous Cosmopolitics in the Andes: Conceptual Reflections beyond 'Politics.'" *Cultural Anthropology* 25 (2): 334–70.

Dodd, Materoa. 2010. "Effects of Industry on Māori Cultural Values: The Case of the Tarawera River." *Indigenous Voices, Indigenous Research*: 53–63. http://hdl.handle.net/10289/4715.

Haraway, Donna J. 2008. *When Species Meet*. Minneapolis: University of Minnesota Press.

Hunt, Sarah. 2014. "Ontologies of Indignity: The Politics of Embodying a Concept." *Cultural Geographies*. 21 (1): 27–32.

Kanewa, Lisa. 2010. "Climate Change Implications for Māori." In *Māori and the Environment: Kaitiaki*, ed.Rachael Selby, Pātaka Moore, and Malcolm Mulholland. Wellington, New Zealand: Huia Press.

Latour Bruno. 2004a. *Politics of Nature: How to Bring the Sciences into Democracy*. Cambridge, MA: Harvard University Press.

———. 2004b. "Whose Cosmos, Whose Cosmopolitics? Comments on the Peace Terms of Ulrich Beck." *Common Knowledge*. 10 (3): 450–92.

Miller. 2008. "Uncle Tasman's Abuses Investigated through Art." *Whakatane Beacon*, March 7.

New Zealand Ministry of Justice. n.d. "Waitangi Tribunal." Accessed 6 May 2016. http://www.justice.govt.nz/treaty-settlements/waitangi-tribunal.

Park, Geoff. 2007. *Theatre Country: Essays on Landscape and Whenua*. Wellington, NZ: Victoria University Press.

Roberts, Mere, Brad Haami, Richard Benton, Terre Satterfield, Melissa L Finucane, Mark Henare, and Manuka Henare. 2004. "Whakapapa as a Māori Mental Construct: Some Implications for the Debate over Genetic Modification of Organisms." *The Contemporary Pacific* 16 (1): 1–28.

Robertson, Natalie. 2011. Interview by Janine Randerson. Personal interview. Mt Albert, Auckland, NZ. February 25.

———. 2007. Natalie Robertson [homepage]. Accessed 15 April 2016. http://natalierobertson.weebly.com.

———. 2008. *Uncle Tasman—The Trembling Current That Scars the Earth*. Three Channel Video projections and sound. Dimensions of installation variable. Collection of the artist, New Zealand.

Ruru, Jacinta. 2004. "A Politically Fuelled Tsunami: The Foreshore/Seabed Controversy in Aotearoa Me Te Wai Pounamu/New Zealand." *The Journal of the Polynesian Society*. 113 (1): 57–72.

Smith, Huhana. 2012a. "Hei Whenua Ora ki Te Hākari: Reinstating the Mauri of Valued Ecosystems: History, Lessons and Experiences from the Hei Whenua Ora ki Te Hākari/Te Hākari Dune Wetland Restoration Project." Lincoln, NZ: Ngā Māramatanga-ā-Papa [Iwi Ecosystem Services] Research Monograph Series No. 9.

———. 2012b. "Review of *Te Hiko Hou*: New Zealand Film Archive Media Gallery, Wellington. *Circuit: Artist Film and Video Aotearoa New Zealand*. Accessed December 20, 2016.

Spivak, Gayatri. 1988. "Can the Subaltern Speak?" In *Marxism and the Interpretation of Culture,* ed. Cary Nelson and Lawrence Grossberg, 271–316. London: Macmillan.

Stengers, Isabelle. 2005. "The Cosmopolitical Proposal." In *Making Things Public: Atmospheres of Democracy,* ed. Bruno Latour and Peter Weibel, 994–1004. Cambridge, MA: MIT Press.

———. 2011. *Cosmopolitics II*. Minneapolis: University of Minnesota Press.

Strathern, Marilyn. 1990. "Artefacts of History: Events and the Interpretation of Images." In *Culture and History in the Pacific*, ed. Jukka Siikala, 24–44. Helsinki, Finland: Finnish Anthropological Society.

Te Ao, Shannon. 2012a. *Untitled (after Rākaihautū)*. Performance Video. Dimensions variable of single wall projection. Collection of the artist, New Zealand.

———. 2012b. Interview by Anna-Marie White. *Puehu: Cultural Dust*. Tumblr. 20 August, 7:05 P.M. http://puehu.tumblr.com/post/29816037192/untitled-after-rakaihautu-2012-shannon-te-ao-.

Todd, Zoe C. 2014. "An Indigenous Feminist's Take on the Ontological Turn: 'Ontology' Is Just Another Word for Colonialism." *Urbane Adventurer: Amiskwacî*. 24 October. http://zoeandthecity.wordpress.com/2014/10/24/an-Indigenous-feminists-take-on-the-ontological-turn-ontology-is-just-another-word-for-colonialism/.

Watson Matthew. 2014. "Derrida, Stengers, Latour, and Subalternist Cosmopolitics." *Theory, Culture & Society*. 31(1): 75–98.

Whatmore, Sarah. 2002. *Hybrid Geographies: Natures Cultures Spaces*. London: Sage Press.

Wright, Laura. 2010. "Diggers, Strangers and Broken Men." *Postcolonial Green: Environmental Politics and World Narratives*, ed. Bonnie Roos and Alex Hunt, 64–79. Charlottesville: University of Virginia Press.

Yates Amanda. *Pop-up Garden/Māra Iti*. 2012. Temporary garden installation. Dimensions Variable. Collection of the artist, Wellington Civic Square. Wellington, New Zealand.

2 Science Fiction, Westerns, and the Vital Cosmo-ethics of *The 6th World*

Salma Monani

> Why science fiction? ... Why not science fiction? Are First Nations people supposed to be portrayed only in Westerns as losers to the greater good of manifest destiny?
>
> —Sierra Adare (2005, 6)

Navajo filmmaker Nanobah Becker's *The 6th World: An Origin Story* (2012) is unmistakably science fiction. Made in collaboration with Independent Television Service's Futurestates series, the short film takes place on a spaceship piloted by scientists on a mission to colonize new worlds. Becker's characters wear insignia-stamped uniforms, high-tech monitors count the number of days to their destination, and doors with vacuum seals whoosh and whir as they open and close. Becker mentions that her own science fiction inspiration was Stanley Kubrick's *2001: A Space Odyssey* (1968), and *The 6th World* reflects that film in its vintage-sixties set design as well as its theme of space travel.

Films like *2001*, in which space is an unexplored frontier, often replicate discourses central to the Western genre (e.g., Abbott 2006; Katerberg 2008) and highlight what historian Carl Abbott reminds us is science fiction's "uneasy companionship" (2006, 12) with the Western. Yet *The 6th World* subverts colonial motifs that tie these genres together; instead, it emphasizes an Indigenous perspective. This perspective is an important addition to our understandings of cinematic genres for a number of reasons, from its revisioning of racialized spaces to its eco-imaginative potentials. Specifically, Becker draws inspiration from Navajo creation myths that describe the migration of Diné people through a series of worlds over cosmic space and time. Given that many Navajo recognize our current world as the fifth in this series of migrations, *The 6th World* speculates on possible futures.

In taking up this futuristic form, the film lends itself to discussions of Indigenous science fiction (sf), or what Grace L. Dillon's seminal *Walking in the Clouds: An Anthology of Indigenous Science Fiction* (2012) broadly conceives of as Indigenous speculative fiction. While focused on literary texts, Dillon's anthology engages extensively with diverse Indigenous literature, including narratives of slipstream and contact, Indigenous scientific literacies, and Native apocalypse. The anthology describes sf as enabling an envisioning of "Native futures, Indigenous hopes, and dreams" (2012, 2).[1] Extending Dillon's literary focus to film, William Lempert's (2014) "Decolonizing Encounters of the Third Kind: Alternative Futuring in Native Science Fiction Film" surveys the recent surge in Indigenous sf film. He briefly touches on *The 6th World* to highlight the usefulness of sf as a genre that encourages imagination of a "multiplicity of futures for their [Indigenous] communities while critically addressing contemporary issues" (Lempert 2014, 164). Similarly, Gabriel Estrada (2014) draws attention to how the film's sf futuring becomes a means to highlight Navajo life-ways and spiritual traditions that are often sidelined by Western colonial and scientific literacies.

Although these scholars laud the film's Indigenous focus, neither interrogates the film's place in the canon of the space travel genre itself. Such an interrogation, I suggest, furthers Estrada and Lempert's explorations of *The 6th World* as it deepens our understanding of the decolonizing potential afforded by Indigenous appropriations of popular genres. Situating *The 6th World* as a response to the (space) Western, I argue that through these kinds of Native futurist imaginings, we re-encounter the frontier as Indigenous home. In turn, I suggest, we might glean useful ecological insights from such interventions.

To elaborate on this central thesis, I break my discussion into three parts. First, to demonstrate how *The 6th World* reclaims frontier space, I draw attention to scholarship that helps us recognize the continuity between sf, which primarily looks forward into the future, and the Western, which usually looks back into the past. Second, I point to *The 6th World*'s intervention into this continuity, drawing particular attention to the film's engagement with signature Navajo themes—the important status of women, land, and corn. Third, I link these themes to Becker's broader ecological imperative—"I'm thinking about global warming and degradation of the planet and about its consequences ... we need to be thinking about how we treat our environment and about how that has to stop; we have to change and do something new" (Becker 2012). I argue that *The 6th World* presents us with the possibilities of "something new." Specifically, in its representation of Navajo relations to corn, it engages what I would like to call a "vital cosmo-ethics" of Indigenous vision.

To articulate these vital cosmo-ethics, I situate my discussion in the framework of cosmopolitics, which recognizes that the "political arena is peopled with shadows of that which does not have, cannot or does not want a political voice" (Stengers 2005, 996). As Isabelle Stengers points out in "The Cosmopolitical Proposal" these shadows are human and more-than-human beings living enmeshed in webs of relationality, that is, in a "cosmos," of multiple agencies (Stengers 2005; see also Adamson 2012a, c, and d; de la Cadena 2010; Monani 2014b).[2] The film's recognition that humans are one amongst many in a complex pluriverse, and that this pluriverse isn't easily knowable but must be respectfully and creatively negotiated makes

The 6th World and other Indigenous films important intermediations not only in re-visioning Indigenous futures but in helping us all imagine alternative ecological pasts, presents, and futures.

From the Western to Science Fiction: Frontiers Past and Future

As Sierra Adare (quoted in the epigraph) alerts us, the Western has mythologized American Indians on screen. Developing out of settler dreams and anxieties, the Western is the quintessential expression of white America. Overlaying personal and national imaginations of frontiers between civilization and savagery, opportunity and adversary, the genre all too often portrays Indigenous peoples of the Americas as savage—both noble and ignoble. As Joanne Hearne notes, such Hollywood narratives often claim historical realism and purport to both entertain *and* educate Americans on frontier history. She explains, "The Western's narrative investment in the chronology of U.S. national origins" enshrines in popular imagination a skewed, nostalgic sense of patriotic "Americanizing" among its settler population, which persists into the present. Unfortunately, "the orientation of the Western to the national past also encodes a national future," through the "American" ideals it celebrates as essential (Hearne 2012, 11). We can extend Hearne's statement to popular sf and its history, drawing on Dillon's recognition of sf's colonial origins in the mid-nineteenth-century. While sf usually presents stories of the future, these were often intertwined in the colonial agenda, "whose main interest was coming to grips with—or negating—the implications of [these] scientific mixes of 'competition, adaptation, race, and destiny'" (Dillon 2012, 2). In particular, the sf *space travel* genre is, as historian Carl Abbott writes, ultimately a story of frontiers. For example, Gene Roddenberry, the creator of *Star Trek* (with its signature opening "Space: The Final Frontier") characterized his science fiction as a "sort of wagon train to the stars" (2006, 23). Kubrick and Arthur C. Clark allegedly toyed with an alternate title for *2001: A Space Odyssey.* They considered calling it *How the Solar System Was Won,* in a direct reference to the 1962 John Ford epic *How the West Was Won* (Wikipedia).

A more recent example of space travel sf that repurposes the Western frontier narrative is James Cameron's *Avatar* (2009). Though set in space, the film replicates the Western tropes of a white, male protagonist encountering a "New World" full of Navi, the Indigenous people of Pandora. However, with its twenty-first century sensibilities, its message is one of environmental justice (EJ) and eco-limits. It forwards what ecocritic Gerry Canavan might describe as a "spaceman economy" instead of a "cowboy economy"—that is, it views the frontier in a post–Frederick Jackson Turner sense, where the "once-open, once-free horizon" of infinite resources that spurred the enterprising cowboy West and which "drove American history, has now slammed shut" (2014, 6). While many critics have lauded *Avatar* for its positive message of eco-limits and EJ (e.g., Adamson 2012a; see also a number of essays in Taylor 2013), they also acknowledge its allegiance to a hegemonic Western narrative form, specifically its romanticized adventure told from the point of a white male (Adamson 2012a; Justice 2013; Monani 2014a; Morton 2014). In effect, such space Western cinema echoes what a number of scholars point out is problematic of the Western genre itself. Like many sympathetic Westerns that valorize the environment and Native peoples (arguably sometimes in complex and complicated ways), such (space) Westerns usually steer away from letting Indigenous people speak for themselves. Instead, Native characters tend to be subsidiary, embodying stock tropes of the ecological Indian, or the vanishing noble savage—primitives living in the past and in simple harmony with nature (Adare 2005; Swain and Vowel 2014).[3]

Because of this cinematic legacy, many Indigenous people have engaged with Westerns as a means of critiquing the dominant culture's legacies of misrepresentation and racial discrimination. Lee Schweninger, drawing on the work of numerous Indigenous film scholars, writes in the Introduction to *Imagic Moments: Indigenous North American Film* that Indigenous filmmakers "are of necessity continually in dialogue not only with mainstream culture in general but very specifically with Hollywood and especially with the Hollywood Western as a genre" (2013, 10). From articles such as Steven Luethold's 1995 "Native American Responses to the Western" to Joanna Hearne's recent monograph, *Native Recognition: Indigenous Cinema and the Western* (2013), many Indigenous studies scholars have explored the Western for its structural, historical, and cultural place in Indigenous imaginaries and resistances (e.g., Gerster 2013; Huhndorf 2001; Raheja 2011; see also essays in Howe, Markowitz, and Cummings 2013).

Much of this Indigenous "talking back" (Marubbio and Buffalohead 2013) to the Western is because of the Western's power to shape national U.S. and—thanks to Hollywood's international reach—global imaginaries of colonial settlement. In *Gunfight at the Eco-Corral: Western Cinema and the Environment* (2012), Murray and Heumann turn to Native-made films like Chris Eyre and Sherman Alexie's *Smoke Signals* (1998) to highlight how Indigenous filmmakers "talk back" by appropriating the Western genre. We can similarly turn to films such as Becker's *The 6th World* to demonstrate how Indigenous filmmakers talk back to (space) Westerns. In doing so, we encounter the decolonialization agenda both in the portrayal of Indigenous people as lead characters and in the portrayal of ecological relations. Such re-imaginings make room for Indigenous agency and "narratives of environmental adaptation" (Murray and Heumann 2012, 197). In a media landscape dominated by Westerns that often homogenize and reify Indigenous identity, the complex nuances of Indigenous sf, such as that of *The 6th World*, speak to the complicated struggles of Native eco-experiences and politics.

The Frontier Revised through Navajo Eyes: Women Astronauts, Monument Valley, and Corn

In describing the traits of most popular science fiction, Abbott (2006) recognizes three fundamental elements: "a storyline or plot revolving around a dilemma or conflict ..., an extrapolation or speculation that makes the plot intellectually interesting," and finally "a backstory, against which the action is placed" (32). *The 6th World* can be understood within this framework. As her first sf venture, Becker crafted her script over a period of three months with Independent Television Series' Futurestates programming producers. ITVS has hosted four seasons of sf shorts as part of the American Public Broadcasting Service's agenda to support independent filmmaking. Recounting the process, Becker explains, "with this film, I had to give Futurestates what they wanted and make us both happy" (Becker 2013). The result is a tight, fifteen-minute storyline revolving around a generic dilemma for sf space travel: Will the crewmembers succeed in their mission to find a replacement planet for an ecologically devastated Earth? In this case, the mission is to safely travel to Mars, with viable corn stock, that can then be used to colonize the planet.

Scholars like Estrada and Lempert and filmmaker Becker herself, point out that what makes *this* plot intellectually interesting is how the mission is accomplished and who is instrumental to its success. Unlike most mainstream sf, which features white men as heroic actors with perhaps a white woman or person of color as a foil, Becker's primary objective is to forward an Indigenous perspective, "because it's definitely something that is lacking, especially in the U.S., where it doesn't really exist in terms of Hollywood cinema and that kind of mainstream media" (Becker 2012). Thus, Becker's protagonist is a female Navajo astronaut, Tazbah Redhouse (played by Navajo actress Jenada Benally) and the entire premise of the journey rests on Navajo creation myths:

> Some people say we are in the 4th World or the 5th World, so when I'm thinking about the film, I'm thinking about the future from a Navajo perspective and this world right now as an end for this time and the beginning for a new time.
>
> (Becker 2012)

If this journey—into a new time and colonizing a new space—echoes a generic frontier, it also rejects the classic EuroAmerican image of the frontier traveler. There is neither a white male law enforcer who is likely above the law himself (such as Captain Kirk of *Star Trek* [1966–1969]), nor a white male cowboy plagued by a deeply individual and personal past (such as Han Solo of *Star Wars* [1977], Malcolm Reynolds of *Firefly* [2002–2003], and Jake Sully of *Avatar* [2009]). Instead, Navajo actors, as well as those of other non-Anglo ethnicities, are in charge. In addition to Benally, the film casts Roger Willie as Redhouse's Navajo supervisor, General Bahe;

African-American Leith Burke as the Corporate Officer at base; and Latino Luis Aldana as Dr. Smith, Redhouse's traveling companion.

Furthermore, *The 6th World*'s reclamation of Monument Valley as its setting interrupts generic landscape tropes and rethinks the purpose and presence of frontier land in (space) Westerns. Most non-Indigenous audiences recognize Monument Valley (*Tsé Bii' Ndzisgai,i* as it is known to the Diné) not by its geographic specificity (it is part of Navajo reservation land on the present-day border between Arizona and Utah, in the Four Corners Region) but as an iconic feature of the landscape in John Ford's mid-twentieth century Hollywood Westerns. *Smithsonian Magazine*'s Tony Perottat writes, "Thanks to Ford, Monument Valley is one of the most familiar landscapes in the United States, yet it remains largely unknown" as a specific place (2010). Much of what we don't know about the valley is because its landscape has served as a *generic* setting for stories of frontier expansion. Ford used the valley as a backdrop for settler conflicts with various tribes—Apache, Comanche, Cheyenne, Arapaho, etc. It didn't matter which tribal group had in fact lived in the valley historically. His purpose was to evoke a national nostalgia for the settler period, when sublime land, once emptied of its Native presence, presented plentiful, open spaces for colonial appropriation. In other films dating from as far back as the 1920s Western *The Vanishing American* (adapted from Zane Grey's novel of the same name) to the twenty-first century's *Mission Impossible II* (2000), *The Lone Ranger* (2013), and *Transformers: Age of Extinction* (2014), the valley has repeatedly been the setting for films that embrace values of personal freedom and opportunity—the individualistic ideals on which the United States was built and that it propagates through its global, neoliberal reach.

Given its ties to the Western, not surprisingly, popular EuroAmerican sf has also appropriated Monument Valley. For example, Kubrick's *2001: Space Odyssey* incorporates color-filtered panoramic images of the valley in its final sequences where its protagonist, Dr. Dave Bowman, psychedelically encounters what some interpret as the alien world. Abbott writes, "Because new planets are often imagined as places of grand vistas and wide-open spaces, the landscape of the West has provided an easy source for sketching the appearance of new places" (2006, 20). Hearne's critique of the Western can be extended to these (space) Westerns: such symbolic abstraction of the valley undermines the history of the area and its Navajo people, who continue to bear the brunt of colonial and neocolonial intrusions on their homeland.

Becker recalls that early on in the project, when she had decided that the film would be about Navajo on Mars, she "immediately knew I had to shoot in Monument Valley" (2013).[4] Her reason, she explains jokingly, was that "it is red, and Mars is the Red Planet." More seriously, Becker adds, "I always look for an opportunity to make location part of my stories," explaining that the location holds material and cultural significance to her people (2013). "We are on our ancestral land. The stories are in the land, the landscape" (Becker, quoted in Estrada 2014, 524).

Figure 2.1 Still from *The 6th World* (2012). Opening sequence with corn foregrounding Monument Valley's red rocks. Source: Courtesy of Nanobah Becker.

The 6th World opens with the nearly inaudible sound of the wind, two pans of the red cliffs, and a long shot of the valley. Cast in red filters, this long shot lingers, evoking Ford's iconic panoramas of the valley. There are no people on screen, and the landscape appears barren. However, Becker up-ends the generic expectation in Westerns of a hostile land waiting to be tamed with a low angle shot showing a field of corn in front of the red rocks. Another cut to a tighter-focused shot of the corn pans downward along the corn stalks to sunbrushed, pollen-heavy flowers being visited by bees. While corn is actually wind-pollinated, Becker's choice of foregrounding the life-giving centrality of corn and its pollen in the landscape is deliberate. It highlights the Navajo community's links to their homelands through their source of life—corn is both the basis of Navajo food sustenance, as well as the heart of their cultural identity and well-being as a people. As religious studies scholar Thomas Raitt writes, corn is a gift, provided by Changing Woman, the sacred matriarch of Navajo people, "who created the Navajo and gave them pollen as a way of living in contact with her" (Raitt 1987, 524).

The opening sequence of the film that foregrounds corn against Monument Valley's red rocks alerts us right away to the film's acknowledgment of corn as a presence central to Navajo spirituality and life. This sequence turns out to be Redhouse's dream. The serene close-ups of corn thriving on the land *precede* imagery of the uninhabitable valley. They reverse the typical EuroAmerican frontier tropes that begin with a sense of the land as uninhabitable, which then move to possibilities of habitation. Here the live landscape shots are followed by a sharp cut to a close-up of red dirt, on which dry, dead corn stalks fall with slapping sounds. As the close-up dissolves to a long shot of a single stalk remaining tall and solitary against the red rocks, wind, then the whisper of Navajo chants fill the soundscape. The chants get louder; golden sparkles emerge, swirl on screen, and hover beside the standing corn as a spirit-being that revitalizes the threatened landscape. Such spirit-beings (*kachina*), which are common in Navajo traditions, serve to represent the natural world, or cosmos. Estrada writes, "Becker carefully took cues from Navajo traditions to guide her to create CGI [computer-generated imagery] effects in the Monument Valley area of Navajo Nation, where she depicted both the Holy People and the Martian landscape" (524). This peopling of the landscape with more than the human presence invites viewers to contemplate Indigenous spiritual relations, which emphasize ecological relations. And, it does so in ways that question not only colonial settler narratives of conquest but also neo-colonial discourses of present-day capitalism.

Figure 2.2 Still from *The 6th World* (2012). Opening sequence depicting a Navajo spirit-being associated with corn. Source: Courtesy of Nanobah Becker.

Specifically, the film pits genetically modified (GMO) corn owned by the mission's company, Omnicorp, against traditional Navajo corn. Estrada (2014) provides an eloquent reading of the film's cinematography to demonstrate how the film critiques "the postmodern, corporate sterility of agribusiness while emphasizing the survival of Navajo Nation migration beliefs, matriarchy, and spirituality" (523). One can further his reading to argue for the film's eco-political significance as many Indigenous peoples struggle for food sovereignty against biotechnology companies such as the film's fictional Omnicorp presence (see, for example, Adamson 2012d).[5] In *The 6th World*, Becker represents this politicized struggle through the mission's restrictive attitude towards Navajo corn. Early in the film, just before take-off, General Bahe, Redhouse's supervisor, attempts to openly gift her with Navajo corn pollen, but Omnicorp's Dr. Smith reprimands him sternly, "Only company-approved corn substances are allowed." The significance of the gift of corn pollen is not lost on Navajo audiences, as it runs through "*every* aspect of Navajo ceremonial life" (Raitt 1987, 525). Nor is the significance of its rejection surprising within the contemporary framework of colonial and neoliberal logics that impose their sense of technological superiority and capitalist monopolies over Indigenous knowledge systems (see for example, Grey and Patel, 2015).

However, *The 6th World* soon turns the tables on such biotechnical logics. The GMO corn stock begins to fail, endangering the mission. In its place, Navajo corn takes primacy, as General Bahe had smuggled Navajo corn onboard in a box bearing the Navajo Nation flag that he presented to Redhouse as a goodbye gift. Redhouse stumbles upon the corn quite by accident at the darkest moments in the film (both figuratively and literally). As the Omnicorp corn stock lies blackened and wasted in its petri-dishes, the mise-en-scène reveals gloom and trouble, with alarms buzzing, Smith slumped in a corner, and a disheveled Redhouse leaning weakly on her desk. Muttering an apology to Bahe for her failed mission, Redhouse hugs the gift box. As she does, she discovers that Bahe has hidden the Navajo corn in the flag. Cinematically both corn and flag serve as beacons of hope, cheerful brightness permeating the gloom of the dark set. Redhouse exclaims to Smith that using her ancestors' corn can help "rebalance the system." Explicit in this assertion is Navajo cultural lore that recognizes traditional corn as life-giving, but one might also speculate on the implicit presence of Indigenous scientific literacies, which can be adapted to the circumstances of the moment—through the careful replanting of this corn, and its successful growth, the reintroduction of photosynthetic processes can feed the ship's oxygen supply and, in turn, ensure the survival of its inhabitants.[6]

The elevation of Navajo literacies helps us grasp at *The 6th World*'s backstory, the Western histories and discourses it challenges. Dismissing Indigenous knowledge systems as primitive superstitions and exploiting America's Indigenous people, including the Navajo, was a project of white men who wrote and broke laws to suit their "cowboy economy" of conquest. Through its lead female Navajo character, whose goal is to sustain, not exploit, *The 6th World* disrupts the primacy of such "cowboy economy" narratives. It also inserts a refreshing counterpart to the "space*man* economy" narrative of popular sf. And it does so by giving agency not just to Redhouse but also to corn, a nonhu*man* entity.

In expanding agency to nonhuman entities, *The 6th World* prods us toward a "cosmo-ethics" that makes room for vitally important alternative ways of seeing and acting, expanded perspectives on space-time horizons, and new eco-political possibilities. As I describe below, by drawing on the rich, growing scholarship in cosmopolitical theory, such cosmo-ethics counter simplistic readings of the film as a reiteration of the stereotypical ecological Indian motif, instead foregrounding a crucial humility needed by humans as we confront the ecological crises of the past, present, and future.

Vital Cosmo-ethics: Careful Hesitations, Flexible Adaptations, and Expanded Horizons

Made most visible by science studies theorists Bruno Latour (2004), Isabelle Stengers (2005), Marisa de la Cadena (2010), and Donna Haraway (2012), cosmopolitics is a conceptual framework that acknowledges more-than-human beings as active agents rather than mere objects outside of human society. To confront the complexities of ecological crises such as climate change, scholars have looked for alternative ways to understand human/more-than-human relations besides the dichotomous or hierarchical models they see as ineffectively proffered by traditional Western scientific literacies. Given cosmopolitics' roots in Indigenous worldviews, many Indigenous studies scholars embrace it as a fruitful concept (Adamson 2012a and c; de la Cadena 2010; Monani 2014a; Poirier 2008; Rose, 2012). Scholars such as Adamson and de la Cadena point to the language of legal documents to articulate cosmopolitics' explicitly productive use in direct Indigenous protest. They and others also turn to Indigenous modes and aesthetics of storytelling—whether oral, written, visual, or otherwise—as these too speak to alternative ways of engaging ecological awareness beyond Western rationalism. In furthering these links to the artistic, Sylvie Poirier's expansion of the term cosmopolitics to "cosmopolitics-poetics" is especially productive:

> In Indigenous socialities, politics and poetics (aesthetics) cannot be disentangled ... the political acts of producing and reproducing the diverse relations between humans and nonhumans—as political acts of alliance and exchange, of communication and negotiation—always imply forms of art and creativity in the sense that they involve aesthetic and performative aspects.
>
> (Poirier 2008, 50)

One can distill Poirier's quote to the recognition of three central cosmo-ethical insights: the necessarily intertwined presence of (a) human and more-than-human, (b) art and politics, and (c) *respectful* "alliance and *exchange*, communication and *negotiation*." In adding the italics, I wish to draw attention to Poirier's point that:

> ... nonhuman beings and agencies express themselves in languages that are not always decipherable or accessible to human beings, hence the importance of mediating practices and performances, and of very complex, flexible, and polysemic sign systems and modes of interpreting and deciphering them, which are always creative acts. (2008, 50)

As Poirier writes, mediation involves "a broad range of ritual practices, performances, and experiences" (2008, 50), including ceremonies, shamanistic practices, and dream practices, which while taken seriously are nonetheless adaptable and "flexible" as humans work to communicate with the cosmos's various beings.

Given that *The 6th World*'s opening sequence is Redhouse's dream, the film's prioritizing of such flexible mediation and negotiation is clear. In showing how Redhouse learns to interpret her dream, *The 6th World* adapts traditional Navajo practices and worldviews to the future in ways that deny hubristic certainty. Redhouse's actions illuminate a vital cosmo-ethical "space for hesitation regarding what it means to say 'good'" (Stengers 2005, 995), or right or certain. Thus, in the plot, Redhouse carefully and respectfully navigates which knowledge system she should prioritize, that of her ancestors or that of Omnicorp's reductionist science. She does not dismiss the latter summarily, but works to flexibly meld its methods into a more holistic practice, as demonstrated in the scenes that follow her discovery of the Navajo corn.

Here, after seeding the discovered corn kernels in the lab, Redhouse and Smith place themselves in the ship's hibernating pods, presumably to draw as little as possible on the ship's vulnerable "unbalanced" systems. In reductionist terms, we can speculate that because hibernation saves oxygen that can be used for the corn's germination, this is a sensible adaptive measure. We can also imagine that as the seeds germinate, their leaves will begin producing oxygen via photosynthesis, thus replenishing the ship's oxygen supply. However, the film looks beyond a simple reductionist approach. As Smith is sealed off in his hibernating pod, he encourages Redhouse to "say a prayer." Once in her own pod, Redhouse fervently begins to chant a prayer.

Her prayer constitutes a recognition that her knowledge doesn't guarantee success, that survival possibilities must be deferentially negotiated. As she sings the Navajo Hoop Song, the film shows her transported to an astral, cosmic plane. Behind her, galaxies rotate, so that she appears to be floating in space. She is dressed in ceremonial clothing and her chants accompany a formalized ritual, involving intonations with yellow and white corn, which she holds during the ceremony. As Estrada (quoting Paula Allen Gunn) writes, the sacred hoop of southwestern tribes refers to "Mother Earth" itself (2014, 527). Literally, Becker has located Redhouse beyond Earth, but in a space also occupied by the spirit-beings (*kachinas*) that co-inhabit earth (e.g., Changing Woman, matriarch and creator of corn).

Via Redhouse's prayer ceremony, Becker pays homage to Navajo practices of "Mother Earth" mediation while demonstrating flexibility in how (or where) such cosmopolitical communication occurs.[7] Further, in respecting the concerns of her community elders regarding what rituals she shared with a general audience (see Estrada 2014, 524), Becker's ethics of engagement with Indigenous traditions emphasize that the larger cosmos is often one of unknowns. Multiple worlds are on screen. The audience is uncertain whether Redhouse is herself or has been transformed into a spirit-being. Estrada, drawing from Becker's own understandings, likens her to White Shell Woman and Changing Woman, both central figures in Navajo creation cosmologies

(2014, 528–29). Corn is essential to the ceremony. While relations between humans and the more-than-human blur, generating a mingling of cosmic matter and light, physicality and astral immateriality, as audience members we are nonetheless left with a sense of the cosmic unknowable, or what ecocritic Tim Morton might describe as the "voids, gaps, and wormholes to other universes" (2013, 216). Thus, when we return to Redhouse, awakening in her space pod and to the film's following utopian ending that cuts to an intertitle "years later," which reveals a successfully established colony on Mars, we might understand that Redhouse's communication was successful, but we ourselves can only speculate on the particulars of exchange. Nor do we know what additional challenges Redhouse and her team undertook to ensure the colony's presence. What we do know is that Redhouse had to mediate her survival in this larger cosmos of beings. To ensure her and humanity's future, she turned to the support of her community—of humans (General Bahe and Dr. Smith) and more-than-humans (the traditional corn, her ancestors and their spirits, as well as the ship itself, and the planet Mars). Each of these acts involved a creative yet respectful acknowledgment of other beings. Of the need to not rush foolhardily on, but for the ability for dialogic communication, to speak *and* to listen, so as to sustain collective living.

Listening, as Indigenous studies and communication scholar Donal Carbaugh (1999) writes, can promote a sense of dwelling in the world, because it encourages being attuned to those around us. When we don't listen, we often find ourselves out of tune with the world, inhabiting "a place without discourse, or a discourse out of its place ... *displaced* or *dislocated*" (Carbaugh 1999, 251; emphasis mine). Becker, recalling the Navajo origin stories that inspired her film, highlights how they draw attention to the ecological ramifications of being in dissonance: "There have been different reasons why the world we were in that moment was no longer inhabitable and usually they [sic] had the view of human beings breaking taboos, or something along those lines, which made the environment hostile for the people" (Becker 2012). Navajo traditions emphasize a need for ecological balance predicated on active and respectful communication with the natural world.

To emphasize her point, in *The 6th World* Becker literally dislocates and displaces Redhouse from the earth onto a spaceship traveling between worlds. To relocate herself, Redhouse must plug in, speak, and listen to the more-than-human community. As Poirier's earlier quote suggests, the knowledge of more-than-human beings is not always easy to decipher, and interpretation, as Carbaugh writes, is "for the listener to decide. The meaning will be the listener's" (Carbaugh 1999, 260). Redhouse's struggles in communicating and the deliberate unknowns of the ceremony scene defy simplistic ideas of Indigenous people as somehow "essentially" knowing nature, or "naturally" being ecological. Instead, they point to what Jace Weaver describes as "the honest and extremely difficult struggles of Indigenous peoples to meet ecological challenges confronting them" (1996, xvi).

Redhouse's dislocation from an old homeland also challenges any essentialist understandings of Indigenous "sense of place." Indigenous people, like all people, must work hard to acquire ecological knowledge. Even as the film articulates Indigenous knowledge as an evolving epistemology rather than an essentialized harmony, its presentation of the respectful cosmo-ethics inherent to this epistemology suggests that we might look to traditional Indigenous knowledge as being more attuned to ecological insights than reductionist Western traditions. As sf, *The 6th World* prompts us to contemplate a literal expanded cosmos, which is peopled by more than humanoid aliens and gives agency to all things, including corn and spectral forms.

In evoking this cosmos, Becker also expands Redhouse's and the audience's imaginary time horizons. She has Redhouse reach into her ancestors' ways to help guide Redhouse even as she melds the future into this timescape. During its fifteen minutes, the film manages to evoke the long spatiotemporal scales that ecocritic Rob Nixon points out are so often sidelined, yet are at the heart of our environmental crises. It is on these spatiotemporal scales that slow violence—"violence that occurs gradually and out of sight, a violence of delayed destruction that is dispersed over time and space, an attritional violence that is typically not viewed as violence at all" (Nixon 2011, 2)—plays out. While (as Nixon and many others write) climate change most severely affects the poor and historically marginalized (including many Indigenous communities), its reach is ultimately planetary. Becker's *The 6th World* is in many ways a post-earth apocalyptic tale—all of humanity is in danger because of the ecological hubris of some.

It is also a post-apocalyptic tale of caution and hope. Just as Redhouse adapts her ancestors' advice to communicate with the more-than-human world, we too might work to attune ourselves to an expanded cosmos's co-existent beings. In doing so, we might be able to make visible and audible those on whom we inflict slow violence and who in turn reach back to us in ways that might be the death of us all. If we rethink our ecological relations as cosmo-ethical, then perhaps we engender a vital survival just as Redhouse does.

Reshaping Genres, Revisioning Worlds

Ultimately, human violence against the more-than-human world and Indigenous peoples occurs at all scales. However, one might argue that representational violence is among the slower, more insidious types of violence, acculturated through years of repetition (e.g., Huhndorf 2001, Hearne 2011). As de la Cadena (2010) points out, centuries of hegemonic discourses of EuroAmerican techno-rationalism representing nature as an apolitical object dismiss the agential knowledge systems of Indigenous cultures as simple and primitive superstitions. The real violence of such discourses manifests itself in debates such as those of GMO agribusiness. While rationalizing GMOs as the technological fix to feed the world, transnational corporations often use questionable ethics in plant breeding and patenting to enrich themselves at the expense of Indigenous communities, whose agro-ecology literacies are pirated even as the environments on which Indigenous communities depend are decimated (see for example, Adamson 2012b). *The 6th World,* through its focus on Navajo corn draws attention to the particular ideological violence of colonial and neocolonial systems that have marginalized Indigenous ecological knowledge systems. By re-centering Navajo cosmo-ethics, the film troubles these problematic systems, suggesting alternative possibilities of Indigenous and ecological survivance.

In addition, the film's insertion into the (space) Western subgenre critiques the persistent representational violence of such cinematic genres. From the films of John Ford to the related colonial anxieties of sf, the Western has stereotyped or erased Indigenous peoples and their ecological ways of knowing. *The 6th World* interrupts the Western and sf's colonial gaze. It is, thus, at the forefront of a surge in recent Indigenous cinematic sf expression. This surge includes the Iñupiat director Andrew Okpeaha MacLean's *Children of the Northern Lights* (2013), which also aired on Futurestates, the translation of *Star Wars* into Navajo (2013), and growing attention to mainstream sf's neglect, marginalization, and stereotyping of Indigenous perspectives in films located in space, as in Molly Swain and Chelsea Vowel's irreverent online podcast, *Métis in Space* (2014). While Becker herself is humble about her film, admitting that "I'm not a sci-fi buff, or anything like that. It's definitely a new frontier for me" (2012), the film earns a place in the canon of sf that, as Carl Abbott writes, has a "growing tendency to acknowledge the *complexity* of the historical Western experience by imagining ambiguous futures" (2006, 2; italics mine). The film's cosmo-ethics ensures ambiguity is a vital part of *The 6th World.* Becker's decision to situate the film around the idea of colonizing a new world lends a crucial complexity to the story and prompts viewers to contemplate the very meaning of colonization, and how we encounter "others." Like the imaginaries of multiple Indigenous writers and filmmakers (for a sampling of each, see Dillon 2012 and Lempert 2014 respectively), as well as those of other traditionally marginalized ethnicities (e.g., writers like Alejendra Morales, Octavio Butler, and Karen Tei Yamashita; and filmmakers like Alex Rivera, Jennifer Phang, and Spike Lee), Becker in *The 6th World* reshapes the boundaries of the sf genre. When coupled with the film's interpretation of the politics of Indigenous corn, such reshaping makes room for righting the representational violence of the past and for re-visioning not just past and future but also their inevitable entanglement in the ecologies of our present. Its cosmo-ethics offers a vision of expanded relations that demand careful and artful political-poetic negotiations as we struggle to forward planetary livability for all.

Acknowledgments

I am especially grateful to Nanobah Becker for generously sharing her thoughts on her film with me and for her immense patience. I would also like to thank the organizers of the ImagineNative Film and Media Festival for enabling me to participate in the festival; it is where I first saw the film and met its filmmaker. I also would like to thank Janet Fiskio, Joni Adamson, and Matt Beehr for their invaluable suggestions and insights on the draft of this chapter.

Notes

1. Dillon provides an overview of these categories in the introduction to her collection (2012, 2–12). For example, stories that discuss "time travel, alternate realities and multiverses" are described as slipstream, whereas stories of contact, though self-explanatory, are usually told not from the point of view of the outsider but from that of Native peoples. Gerald Vizenor's 1978 "Custer on the Slipstream" serves as a nice example for both categories and helps forward Dillon's point that "Indigenous sf is not so new—just overlooked" (2012, 2).
2. In my use of the term "vital cosmo-ethics" to consider such agencies, one might also see reference to political theorist Jane Bennett's notion of "vital materialism," which similarly recognizes an agential world beyond that of the human—"the capacity of things ... to act as quasi agents or forces with trajectories, propensities, and tendencies of their own" (2009, viii). While there are similarities between Bennett's concept and the vital cosmo-ethics I describe here, primarily the affiliated end goal of ecological awareness and respect that can frame our ethical obligation to the more-than-human world, there are also key differences. Specifically, Navajo cosmo-ethics generally designate matter as subjects, often with theistic characteristics (spirit-beings), while Bennett does not go so far, instead seeing matter in Latour's term as "actants"—neither subject nor object, but something in-between (2009, 8–10)—and as "assemblages" (adapted from Felix Gauttari and Gilles Delueze's term, 2009, 23–24) in thoroughly "nontheistic presentation" (2009, 17). Because the means by which Indigenous knowledge systems prompt ethical cosmological relationships are distinct, and because Bennett's conception draws primarily from Western traditions, I don't draw from Bennett's work as I do from the works of scholars such as those cited here who do speak to the particularities of Indigenous engagements with the cosmos.
3. For an overview of Indigenous studies literature that identities this trope, see, for example, Adamson 2012b and Monani 2014a.
4. Becker is not the first Navajo filmmaker to reclaim the valley for her people. Jeff Spitz and Bennie Klein's award winning *Return of Navajo Boy* (2000) focuses on the communities that call this area home to tell a historically specific story of the illegal uranium mining that was occurring at the height of John Ford's filmmaking career in the 1950s. For an excellent discussion of the film, see Hearne (2010, 198–206). Like Spitz and Klein, Becker's lens refocuses our encounter with Monument Valley, using it as the landscape of her film's Mars.
5. Much has been written about struggles for food sovereignty. Two particularly important voices are those of Gary Paul Nabhan (n.d.) and Vandana Shiva (n.d.), whose many works highlight how contemporary corporate control results in a loss of both traditional biodiversity and cultural life-ways.

6. One might also speculate that the dying GMO corn, no longer able to photosynthesize and thus release oxygen into the air, has thrown the ship's air supply out of whack, affecting all life on board.
7. Thanks to Joni Adamson for alerting me to Brenda Norell's 1 July 2004 *Indian Country Today* article, "Traditional Navajo Farming Songs of Corn and Life," which discusses traditional planting of corn by the position of the stars. *The 6th World*'s astral ceremony here can be read as respecting such cosmic alignments.

References

Abbott, Carl. 2006. *Frontiers Past and Future: Science Fiction and the American West.* Lawrence: University Press of Kansas.

Adamson, Joni. 2012a "Indigenous Literatures, Multinaturalism, and *Avatar*: The Emergence of Indigenous Cosmopolitics." *American Literary History* 24 (1): 143–62.

______. 2012b. "Seeking the Corn Mother: Transnational Indigenous Organizing and Food Sovereignty in Native North American Literature." In *Indigenous Rights in the Age of the UN Declaration*, ed. Elvira Pulitano, 228–49. Cambridge, UK: Cambridge University Press.

———. 2012c. "'¡Todos Somos Indios!' Revolutionary Imagination, Alternative Modernity, and Transnational Organizing in the Work of Silko, Tamez, and Anzaldúa." *Journal of Transnational American Studies* 4 (1): 1–26.

———. 2012d. "Whale as Cosmos: Multispecies Ethnography and Contemporary Indigenous Cosmopolitics." *Revista Canaria de Estudios Ingleses*, 64 (1): 29–45.

Adare, Sierra. 2005. *"Indian" Stereotypes in TV Science Fiction: First Nations' Voices Speak Out.* Austin: University of Texas Press.

Becker, Nanobah. 2012. Personal (face-to-face) interview. Toronto, Ontario, 19 October.

———. 2013. Skype Interview: Los Angeles. 01 November.

Bennett, Jane. 2010. *Vibrant Matter: A Political Ecology of Things.* Durham, NC: Duke University Press.

Canavan, Gerry. 2014. "Introduction: If This Goes On." In *Green Planets:Ecology and Science Fiction*, ed. Gerry Canavan and Kim Stanley Robinson, 1–24. Middletown, CT: Wesleyan University Press.

Carbaugh, Donal. 1999. "'Just Listen': 'Listening' and Landscape among the Blackfeet." *Western Journal of Communication* 63 (3): 250–70.

de la Cadena, Marisol. 2010. "Indigenous Cosmopolitics in the Andes: Conceptual Reflections beyond 'Politics.'" *Cultural Anthropology* 25 (2): 334–70.

Dillon, Grace, L., ed. 2012. *Walking the Clouds: An Anthology of Indigenous Science Fiction.* Tucson: University of Arizona Press.

Estrada, Gabriel. 2014. "Navajo Sci-Fi Film: Matriarchal Visual Sovereignty in Nanobah Becker's *The 6th World*." *Journal of the American Academy of Religion* 82 (2): 521–30.

Gerster, Carole. 2013. "Native Resistance to Hollywood's Persistence of Vision: Teaching Films about Contemporary American Indians." In *Native Americans on Film: Conversations, Teaching, and Theory*, ed. M. E. Marubbio and Eric L. Buffalohead, 141–74. Lexington: University Press of Kentucky.

Grey, Sam, and Raj Patel. 2015. "Food Sovereignty as Decolonization: Some Contributions from Indigenous Movements to Food System and Development Politics." *Agricultural and Human Values*. 32 (3): 431–44.

Haraway, Donna. 2012. "Cosmopolitical Critters: Companion Species, SF, and Staying in Trouble." John Coffin Memorial Lecture. Cosmopolitan Animals Conference, University of London. London, UK, 26–27 October.

Hearne, Joanna. 2012. *Native Recognition: Indigenous Cinema and the Western.* Albany, NY: SUNY Press.

Howe, LeAnne, Harvey Markowitz, and Denise Cummings. 2013. *Seeing Red: Hollywood's Pixelated Skins. American Indians and Film.* Lansing: Michigan State University Press.

Huhndorf, Shari M. 2001. *Going Native: Indians in American Cultural Imagination.* Ithaca, NY: Cornell University Press.

Justice, Daniel Heath. 2013. "Afterword: Considering the Legacies of *Avatar*." In *Avatar and Nature Spirituality,* ed. Bron Taylor, 334–50. Waterloo, Ontario: Wilfrid Laurier University Press.

Katerberg, William H. 2008. *Future West: Utopia and Apocalypse in Frontier Science Fiction*, Lawrence: University Press of Kansas.

Latour, Bruno. 2004. "Whose Cosmos? Which Cosmopolitics? Comments on the Peace Terms of Ulrich Beck." *Common Knowledges*. 10 (3): 450–62.

Lempert, William. 2014. "Decolonizing Encounters of the Third Kind: Alternative Futuring in Native Science Fiction Film." *Visual Anthropology Review* 30 (2): 164–76.

Leuthold, Steven M. 1995. "Native American Responses to the Western." *American Indian Culture and Research Journal* 19 (1): 153–89.

Marubbio, M. E., and Eric L. Buffalohead. 2013. *Native Americans on Film: Conversations, Teaching, and Theory*. Lexington: University Press of Kentucky.

Monani, Salma. 2014a. "Evoking Sympathy and Empathy: The Ecological Indian and Indigenous Eco-activism." In *Moving Environments: Affect, Emotion, Ecology and Film,* ed. Alexa Wiek von Mossner, 225–47. Waterloo, Ontario, Canada: Wilfrid Laurier University Press.

———. 2014b. "*Kissed by Lightning* and Indigenous Cinema's Natureculture Continuum." In *Ecoambiguity, Community and Development*, ed. Swarnalatha Rangarajan, Vidya Sarveswaran, and Scott Slovic. 131–47. New York: Lexington Press.

Morton, Timothy. 2014. "Pandora's Box: Avatar, Ecology, Thought." In *Green Planets: Ecology and Science Fiction*, ed. Gerry Canavan and Kim Stanley Robinson, 206–25. Middletown, CT: Wesleyan University Press.

Murray, Robin L., and Joseph K. Heumann. 2012. *Gunfight at the Eco-Corral: Western Cinema and the Environment.* Norman: University of Oklahoma Press.

Nabhan, Gary Paul. n.d. *From the Field, to the Campfire, to the Kitchen* (blog). http://garynabhan.com/i/.

Nixon, Rob. 2011. *Slow Violence and the Environmentalism of the Poor.* Cambridge, MA: Harvard University Press.

Perottat, Tony. 2010. "Behind the Scenes at Monument Valley." *Smithsonian Magazine*. February. http://www.smithsonianmag.com/travel/behind-the-scenes-in-monument-valley-4791660/?no-ist.

Poirier, Sylvie. 2008. "Reflections on Indigenous Cosmopolitics-Poetics." *Anthropologica* (1): 75–85.

Raheja, Michelle. 2010. *Reservation Reelism: Redfacing, Visual Sovereignity, and Representations of Native Americans in Film*. Lincoln: University of Nebraska Press.

Raitt, Thomas M. 1987. "The Ritual Meaning of Corn Pollen among the Navajo Indians." *Religious Studies*, 23 (4): 523–30.

Rose, Deborah Bird. 2012. "Cosmopolitics: The Kiss of Life." *New Formations*. 76 (3): 101–13.

Schweninger, Lee. 2013. *Imagic Moments: Indigenous North American Film*. Athens: University of Georgia Press.

Shiva, Vandana. n.d. *Navdanya*. (website). http://www.navdanya.org/home.

Stengers, Isabelle. 2005. "The Cosmopolitical Proposal." In *Making Things Public: Atmospheres of Democracy*, ed. Bruno Latour and Peter Weibel, 994–1004. Cambridge, MA: MIT Press.

Swain, Molly, and Chelsea Vowel. 2014. *Métis in Space*. Podcast audio. Accessed July 22, 2015. http://www.indianandcowboy.com/metis-in-space-podcast/.

Taylor, Bron. 2013. *Avatar and Nature Spirituality*. Waterloo, Ontario: Wilfrid Laurier University Press.

Vizenor, Gerald. "Custer on the Slipstream." In *Walking the Clouds: An Anthology of Indigenous Science Fiction*, ed. Grace L. Dillon, 15–26. Tucson: University of Arizona Press.

Weaver, Jace. 1996. Preface to *Defending Mother Earth: Native American Perspectives on Environmental Justice*, ed. Jace Weaver, xv–xvi. Maryknoll, NY: Orbis Books.

Wikipedia. s.v. "2001: A Space Odyssey (film)," last modified 25 April 2016. Accessed 2 May 2016. https://en.wikipedia.org/wiki/2001:_A_Space_Odyssey_(film).

Films

Avatar. 2010. Directed by James Cameron. Beverly Hills, CA: 20th Century Fox.

Children of the Northern Lights. 2013. Directed by Andrew Okpeaha MacLean. San Francisco, CA: Independent Television Service, Inc. (ITVS).

Firefly 2002–2003. Directed by Josh Whedon. Beverly Hils, CA: Mutant Enemy Productions.

Return of Navajo Boy, The. 2011. Directed by Jeff Spitz and Bennie Klain. Chicago, IL: Groundswell Educational Films.

6th World, The: An Origin Story. 2012. Directed by Nanobah Becker. San Francisco, CA: Independent Television Service, Inc. (ITVS).

Smoke Signals. 1999. Directed by Chris Eyre and Sherman Alexie. Montreal, Quebec: Alliance Vivafilm.

Star Trek 1966–1969. Produced by Gene Roddenberry. Hollywood, CA: Paramount Television and NBC.

Star Wars. 1977. Directed by George Lucas. Beverly Hills, CA: Twentieth Century Fox.

Star Wars: A New Hope. Episode IV. 2013. Translated into Navajo. Beverly Hills, CA: Twentieth Century Fox and Navajo National Museum.

3 Long Environmentalism

After the Listening Session

Subhankar Banerjee

In August 2006, the U.S. Secretary of the Interior, Dirk Kempthorne, went to Fairbanks, Alaska, to listen to the people talk about their views on energy development in the Arctic. Fairbanks was one of the several stops on a nationwide "Listening Session" tour by the Secretary. After hearing overwhelming opposition to the proposed expansion of oil and gas development in the Alaskan Arctic from the people gathered in a large auditorium, the Secretary told the audience (I am paraphrasing from memory): "I have listened to your concerns, but I must follow the President's mandate to open up the Arctic land and water to oil and gas development." The conservationists felt hoodwinked that day. To remember the farcical nature of the event, I made a photograph, *After the Listening Session*. The picture is a group portrait of nine individuals, all of whom have been engaged in protecting significant biocultural areas in the Alaskan Arctic from industrial exploitation. Historically, environmentalists and Indigenous peoples were not traditional allies, as I will soon explain, nor were the Gwich'in and the Iñupiat peoples of Arctic North America. The photograph *After the Listening Session*, however, holds all of them together. The picture opens a doorway into what I call—*long environmentalism.*[1]

When an environmental engagement has lasted for a while—say a quarter-century, or more—it creates a culture of its own, has its own history. Such a multiple-decades-long engagement gives rise to its own distinct form of environmentalism, or what I will call in this chapter "long environmentalism." The two principal tenets of long environmentalism include: collaboration among unlikely allies through the act of sincere listening, giving rise to *radical hope*; and a period of time that is long enough to enable what was once considered marginal (like a human community or an idea) to become significant and essential. The unlikely allies could be historical adversaries or groups of people who come from different cultures, races, classes, and geographies. The coming-together of unlikely allies, however, creates varieties of contradictions, which often cannot be resolved but will have to be held in place. While holding varieties of contradictions in place within a single engagement, long environmentalism performs four related functions: it illuminates past injustices, highlights the significance of resistance movements to avert potential social-environmental violence (fast and/or slow), shows how communities respond to slow violence, and points toward social-ecological renewal after devastation. In doing so, it gives radical hope to ideas of coalitional politics, where coalitions are forged with the ethics of livability that pay attention to human and nonhuman alike. I suggest that long environmentalism is an ethic that is universally relevant yet meets cultural practices that are situated in the local.

A study of long environmentalism begins with a statement that specifically addresses multiple-decades-long engagement. Here is one example, which I will return to: "We must fight and do all we can to preserve our way of life even if we feel like we have been fighting the same fight for the last *fifty years*" (Cannon 2013, 320; emphasis mine). Long environmentalism, when engaged for the purposes of an ecocritical study, would analyze such statements, fleshing out their historical, sociopolitical, and ecological contexts.

In this chapter I analyze two case studies from post-1950 Arctic Alaska, that of the Arctic National Wildlife Refuge located in Northeast Alaska and that of the Beaufort and Chukchi seas of the Arctic Ocean, for the ways they illustrate long environmentalism. Both are informed in part by my own fieldwork there since 2001. By focusing on the environmental politics of these two case studies, with a history of over half a century for each, I demonstrate how resistance movements of the Indigenous peoples, in this case the Gwich'in and the Iñupiat, help draw the attention of conservationists to issues of environmental justice that redress past histories and illuminate the key tenets of long environmentalism.

Rights of Nature or Environmentalism of the Poor?

Distinguished environmental historian Roderick Nash lays out centuries of Western intellectual evolution in environmental thought in Europe and North America in *The Rights of Nature: A History of Environmental Ethics*, first published in 1989. But what does "rights of nature" mean? Based on his reading of the history of environmental ethics, Nash writes that, "[N]ature has intrinsic value and consequently possesses at least the right to exist. This position is sometimes called 'biocentricism,' 'ecological egalitarianism,' or 'deep ecology,' and it accords nature ethical status at least equal to that of humans" (Nash 1989, 9–10). The principal tenet of deep ecology is biocentricism, meaning that nonhuman biotic life has a right to exist and flourish independent of human intervention and needs. In 2015, a team of scientists announced in a study that "a sixth mass extinction is already underway," that "the average rate of vertebrate species loss over the last century is up to 100 times higher than the background rate," and that "our global society has started to destroy species of other organisms at an accelerating rate, initiating a mass extinction episode unparalleled for 65 million years" (Ceballos et al. 2015). At such a time of epic biological crisis, in what is being termed by many as a new geologic age (the Anthropocene), an ethic that is biocentric is to be welcomed, as it is urgently needed.

The same year Nash's book was published, however, eminent Indian sociologist-turned-historian Ramachandra Guha wrote a combative essay to critique deep ecology, first published in the journal *Environmental Ethics* and later widely anthologized. Guha feared that, "the social consequences of putting deep ecology into practice on a worldwide basis (what its practitioners are aiming for) are very grave indeed" (Guha 1989, 72). He further charged that, "invoking the bogy of anthropocentricism is at best irrelevant and at worst a dangerous obfuscation," while pointing out two key issues that the deep ecologists have remained silent about and what he thought are the key drivers of ecological destruction: overconsumption and growing militarization (Guha 1989, 74). A few years later, in 1994, he introduced, "environmentalism of the poor," considered to be a radical notion at the time, at a conference in Ecuador, and later wrote about it in a co-authored book, *Varieties of Environmentalism: Essays North and South* (Guha and Martinez-Alier 1997). Deep ecology is an ethic. Environmentalism of the poor, on the other hand, is a practice. Guha envisioned it as a combination of "struggles against environmental degradation" and "struggles for environmental renewal" (Guha 2000, 104). He presented it in part as a protest against what he perceived to be an arrogant and false belief held by many mainstream environmentalists in the U.S. that poor people couldn't afford to be environmental. The phrase "environmentalism of the poor" has resonated with other scholars since Guha introduced it. Juan Martinez-Alier published his own book, *The Environmentalism of the Poor: A Study of Ecological Conflicts and Valuation* (2003), and eight years later the phrase appeared in the title of Rob Nixon's postcolonial environmental literary studies treatise, *Slow Violence and the Environmentalism of the Poor* (2011). The rights of nature ethic as articulated by Nash and the environmentalism of the poor practice as articulated by Guha remain, in mainstream environmental politics, uneasy adversaries. In North America, the formalization of the environmental justice movement in the 1990s represents some of this debate and unease, but it also points to grassroots and legal attempts that seek to reconcile concerns for human rights with the survival of nonhuman biotic life (see, for example, Adamson, 2001; Adamson, Evans, and Stein, 2002).

Building on environmental justice discourse that recognizes the interdependency of human and nonhuman, in recent years there has been a new invocation of rights of nature, which was primarily initiated by Indigenous peoples in Latin America. Following the failure of the 2009 United Nations Climate Change Conference (the 15th session of the Conference of the Parties [COP15] to the U.N. Framework Convention on Climate Change) in Copenhagen, Bolivia hosted a World People's Conference on the Rights of Mother Earth and Climate Change in 2010, which was attended by nearly 30,000 delegates from 100 countries. The following year, Bolivia passed the "Law of Mother Earth." The rights of nature have also been discussed in the context of contemporary art and environmental scholarship. Two notable examples from the year 2015 are the *Rights of Nature: Art and Ecology in the Americas* exhibition at the Nottingham Contemporary in the United Kingdom and the associated international conference there (Demos 2015) and the *Conflict Shorelines: History, Politics, and Climate Change* conference at Princeton University, which included a panel, "The Rights of Nature."[2] The contemporary rights of nature discourse has developed within a juridical framework and can be seen as a continuation of, and as consistent with, the long environmentalism in Arctic Alaska, which I will now discuss.

(Un)Inhabiting America's Wilderness: The Alaska National Wildlife *Range*

One of the two braids of long environmentalism embedded in the *After the Listening Session* photograph addresses the resistance movement to protect the Arctic National Wildlife Refuge in northeastern Alaska from industrial incursion. The Arctic National Wildlife Refuge, considered to be the most biodiverse conservation area in the circumpolar North, is one of the most debated public lands in U.S. history, as it harbors some of America's oil and gas reserves. It also happens to provide nutritional, cultural, and spiritual sustenance to two Indigenous communities: the Gwich'in, who live on the south side of the Brooks Range Mountains, and the Iñupiat, who live on Barter Island, along the Arctic coast (Banerjee 2003; Dunaway 2009). The debate over whether to open up the coastal plain to oil and gas development or to protect it permanently has been raging in the halls of the U.S. Congress for nearly four decades. The coastal plain is considered to be the biological heart of the Refuge by scientists and to be *Iizhik Gwats'an Gwandaii Goodlit* ("Sacred Place Where Life Begins") by the Gwich'in people. After five months of background research, in March 2001, I traveled to the Arctic National Wildlife Refuge and spent fourteen months over a period of two consecutive years there, becoming involved in the campaign to protect the Refuge.

On seeing my photograph of Charlie Swaney from Arctic Village scanning the land for animals from his hunting camp in the Arctic National Wildlife Refuge, a young environmentalist asked me during a lobbying campaign in Washington, DC, in 2002 with honest bewilderment: "How could there be a hunting camp in a pristine wilderness?" That question, more than anything else, prompted me to learn about the history of American land conservation.

I was led to the groundbreaking book of microenvironmental history, *Crimes Against Nature: Squatters, Poachers, Thieves and the Hidden History of American Conservation*, by Karl Jacoby (2001). In the nineteenth century, Jacoby points out, when the land conservation movement began to take shape, subsistence hunters—Native Americans and rural whites—were labeled "poachers," inhabitants as "squatters," and subsistence gatherers as

Figure 3.1 Scanning for animals from a Gwich'in hunting/fishing camp, Arctic National Wildlife Refuge near Arctic Village, Alaska. August 2002. Source: Courtesy of Subhankar Banerjee.

"thieves," and those who would set fires for ecocultural reasons as "arsonists," if their homelands were deemed worthy of conservation, effectively criminalizing these subsistence and traditional activities. One of the case studies in the book illuminates the conflict of conservation and Indigenous habitation and land use during the creation of Yellowstone National Park, the first National Park, which was established in 1872. Jacoby points out that five tribes—the Crow, Bannock, Shoshone, Blackfeet, and Nez Perce—actively used the Yellowstone Plateau for subsistence hunting and gathering and that, according to a U.S. Army Corps engineer, "Indian trails ... were everywhere" (83). He then writes that "park backers nonetheless persisted in describing the region as existing in 'primeval solitude,' filled with countless locations that 'have never been trodden by human footsteps'" (84). A bit later Jacoby writes, "Drawing upon a familiar vocabulary of discovery and exploration, the authors of the early accounts of the Yellowstone region literally wrote Indians out of the landscape, erasing Indian claims by reclassifying inhabited territory as empty wilderness" (85). Moreover, the U.S. military managed the park for thirty-two years continuously after its founding to protect the white tourists from the perceived threat of the Native Americans. In the U.S., militarization marked the dawn of land conservation.

Nearly a century later, on December 6, 1960, the U.S. Secretary of Interior Fred A. Seaton signed Public Land Order 2214 (U.S. Fish and Wildlife Service 1960), establishing some of the region now under contestation in northeast Alaska as the Arctic National Wildlife Range. The land was set aside for "the purpose of preserving unique wildlife, wilderness, and recreational values." The first two goals are primarily shaped by deep ecology, but the juxtaposition of all three creates varieties of contradictions when practiced in a single geography. The most problematic part of the Public Land Order, of course, was what it did not include—the Gwich'in and the Iñupiat communities who have inhabited the region for many millennia. This omission, which involves issues of race, class, and gender, was and continues to be characteristic of mainstream wilderness understandings.

Until a few decades ago people of the lower latitudes in North America primarily came to learn about the Arctic from the tales of heroic journeys of white, male explorers. In *Gender On Ice: American Ideologies of Polar Expeditions*, historian of visual culture Lisa Bloom critiqued these gendered-expeditions, and highlighted elements of racism and the significance of nationalism and colonial dominance in those adventures (Bloom 1993). Today, Arctic Indigenous women—activists, artists, and writers—are reclaiming the narratives about their homelands. Gwich'in activist Sarah James, Iñupiaq activist Caroline Cannon, Inuit activist Sheila Watt-Cloutier, Inuit artist Annie Pootoogook, and Gwich'in writer Velma Wallis, to name just a few, are exemplary figures in correcting the white, male-dominated view of the circumpolar North.

Sarah James said, "I learned from living out in the wilderness, our natural world" (James 2013, 260). This is a confrontational statement for American land conservation as, historically, no habitation is permitted inside a designated wilderness. What did James mean by that statement? There are at least two possible interpretations: one literal and the other ethical. Sarah James grew up with her family on the *Sheenjek*, or Salmon River valley, until they were encouraged to take up village life in Arctic Village in the mid–twentieth century. Today, about 150 residents live in Arctic Village, along the East Fork of the Chandalar River. In 2007, during a cold January morning when I visited her home in Arctic Village, she showed me a hand-drawn map of the Sheenjek River valley, with various Gwich'in family camps marked, and lamented the fact that that particular history of Gwich'in habitation and use along the river was obliterated when the Arctic National Wildlife Range was established in 1960.

Figure 3.2 Sheenjek River valley, Arctic National Wildlife Refuge, Alaska. May 2002. Source: Courtesy of Subhankar Banerjee.

Today, that river valley, where James grew up, is a designated wilderness. So in a literal sense she indeed "learned by living out in" what is today a "wilderness." But there is also an ethical interpretation that says that wilderness areas in the Arctic ought not to exclude Indigenous peoples, whose cultural and material subsistence critically depends on these lands and its resources. Not acknowledging habitation of the wilderness had been the case with the formation of the Arctic National Wildlife Range. Yet, as I became more and more engaged in the Arctic Refuge campaign and began to learn about its history, I came to realize that long environmentalism as practiced by people like James sustains itself over a long period of time and is able to accommodate multiple perspectives, in large part through the act of sincere listening—unlike the insincere listening that I witnessed at the Fairbanks Listening Session.

Forging of Unlikely Allies: The Arctic National Wildlife Refuge and Beyond

Given that the Public Land Order 2214 did not recognize the presence of and the traditional rights of the Indigenous communities living in or near the Arctic National Wildlife Refuge, how did the Gwich'in people establish their rights to continue their subsistence way of life? Tlingit scholar Maria Williams has pointed out that Native solidarity, combined with a culture of resistance, eventually succeeded in getting social rights recognized in Alaska, including subsistence rights (Williams 2009).

The passage of the 1980 Alaska National Interest Land Conservation Act (ANILCA) is likely the most significant achievement in U.S. conservation history, as it protected 104 million acres of public lands and waters in Alaska, including significant areas in the Arctic. Following on the heels of the 1970s Alaska Native Claims Settlement Act (ANCSA), it was negotiated with active input from Alaska's Indigenous communities. Thus, while it set aside land for biotic life, it also did something unprecedented—it protected the way of life of the Indigenous communities by granting subsistence rights inside federally protected lands, including wilderness. ANILCA also doubled the size of the original Arctic National Wildlife Range, renaming it a Refuge, and designated eight million acres of it as wilderness where Indigenous people once more had the legal right to subsistence practices. However, even though ANILCA protected a significant part of the Arctic National Wildlife Refuge as wilderness, the coastal plain was left in limbo for the U.S. Congress to decide in future, whether to allow oil drilling there or not.

During the mid-1980s, when the Reagan administration made a push to open up the coastal plain to drilling, the Gwich'in Nation called a historic gathering in Arctic Village in 1988. The community members from fourteen villages in northeast Alaska, as well as the Yukon and Northwest Territories in Arctic Canada, attended the gathering and passed a resolution, the "Gwich'in Niintsyaa" ("Gwich'in Elders Statement" 1988). The resolution called on the U.S. government to recognize the rights of the Gwich'in people and to prohibit drilling in the calving and the post-calving aggregation grounds of the caribou. The Gwich'in saw the drilling as an affront to their material subsistence and to the sacred life of their nonhuman relatives, symbolized most powerfully through their own identification as the "Caribou People." The resolution urged permanent protection of those lands by designating them as "wilderness." With ANILCA's passage, they understood that such a designation would protect the calving grounds of the caribou from commercial exploitation, while maintaining subsistence activities within the Refuge lands.

The 1988 gathering resulted in the founding of the Gwich'in Steering Committee, which continues to work actively with various environmental organizations for the continued protection of the Arctic Refuge, even though they have not always prioritized the same reasons; in particular, the importance of cultural preservation was one that the Gwich'in added to the agenda of conservationists focused on land preservation. This has been a hard-fought coalition, but one that is proving to be effective. The Arctic National Wildlife Refuge continues to remain free of industrial development, while the campaign to protect it permanently, goes on.

Today, we hear about alliances for most contemporary environmental struggles—environmentalists are collaborating with Indigenous peoples, scientists are collaborating with religious leaders. One of the most public examples of an environmental collaboration is Pope Francis's encyclical, *Laudato Si': On Care for Our Common Home* (Francis 2015). A symposium on science, physical and social, at the Vatican in 2014 helped in part lay the foundation for the encyclical. At the same time, the encyclical is also informed by recent scholarship in environmental humanities. Chapter 4 of the encyclical, "Integral Ecology" brings to mind Sean Esbjörn-Hargens and Michael Zimmerman's book, *Integral Ecology: Uniting Multiple Perspectives on the Natural World* (2009). In a critical appreciation of the encyclical, Zimmerman points out that, starting in the 1990s, environmental humanists, including ecocritics, and environmental justice activists have addressed the need to close the traditional Western dualist gap between human and the nonhuman world (Zimmerman 2015). Long environmentalism is a continuation of and contribution to this ongoing multiperspectival effort in environmental humanities, and one of its principal tenets is collaboration between unlikely allies.

While science, especially in modernity, has been suspicious of religion (and vice versa), as we see in the Pope's encyclical, they can work together. In Indigenous traditions, the two have often worked side by side. As this volume attests to, a growing number of environmental humanists, including ecocritics who are writing on "cosmopolitics" and "multispecies ethnography," have highlighted the significance of Indigenous stories as they are being placed on the political stage for the purpose of supportive, innovative alliances for change. Joni Adamson has observed that these stories are being "employed as authoritative commentary/theory illuminating the consequences of global economic development for local humans, animals, and nonhumans" (Adamson 2013, 173).

In the case of Alaska, *Protect: Caribou & Salmon*, a communal and artistic performance by the Gwich'in Nation helps us understand the co-existence of an Indigenous creation myth with science and its significance for social-environmental activism. Alarmed by the devastation caused by British Petroleum's *Deepwater Horizon* blowout in the Gulf of Mexico in April 2010, the Gwich'in Nation gathered in Fort Yukon, Alaska. On July 21, they created *Protect: Caribou & Salmon* on the sandy shore of the mighty Yukon River, to send a message to the world—protect the habitats of the caribou and the salmon, and consequently the culture of "the fish people, the caribou people, or just the Gwich'in people," as Gwich'in novelist Velma Wallis puts it (Wallis 2013, 498). From afar it looks like colorful ants have created three shapes on the sandy shore of a river: the word PROTECT, the outline of an antler, and the outline of a fish. But as we get close we see that the shapes are built out of human bodies, both men and women of all ages, from children to elders. In another era, this communal act would not be considered art, but perhaps part of a spiritual ceremony. In our time, however, the people of the Gwich'in Nation, who participated in the performance, insist that it is *art*—human aerial art. But, protect what? Protect from what? Protect for whom?

In the image, the antler refers to caribou, more specifically the caribou of the Porcupine River herd. That the Gwich'in people created the caribou antler image with their bodies could be understood within the context of their creation story:

> Our creation story tells of the time when there was only animals, the animals became people, when that happened the Gwich'in came from the caribou. There was an agreement between the two that still stands, the Gwich'in retain a piece of the caribou heart and the caribou retain a piece of the Gwich'in heart for all time. We are like one. Whatever befalls the caribou will befall the Gwich'in.
>
> (Gemmill 2001, 49–50)

In the campaign to protect the Arctic National Wildlife Refuge, the environmental organizations speak about science, while the Gwich'in people speak about their creation story. The two views have co-existed peacefully in the fight against a common foe—the fossil fuels industry that wants to drill in the calving grounds of the caribou. The fact that religion and science have co-existed peacefully is radical indeed and has given rise to hope, radical hope for continued survival of the caribou and the eco-spirituality of the Gwich'in that depend on the caribou.

In *Protect: Caribou & Salmon* the salmon are the majestic Yukon River kings, or Chinook that the Gwich'in communities that live along the Yukon River depend on for sustenance. Their presence in the act defies a proposal to open up the Yukon Flats National Wildlife Refuge (which abuts the Arctic National Wildlife Refuge) to oil and gas development. Despite persistent pressure from industry and pro-development politicians, both the Arctic National Wildlife Refuge and the Yukon Flats National Wildlife Refuge continue to remain free of fossil fuel development. In this campaign, the Gwich'in creation story has been used effectively as a statement of resistance against industrial destruction: a myth of the past establishes Indigenous rights in the present and becomes the central argument in the fight for a healthy future—for the human and the nonhuman biotic life.

Indigenous peoples all over the world are experiencing destruction, or the prospect of destruction—of their homelands, food, and culture—from ever-expanding resource exploitation for oil, coal, gas, minerals, timber—to satisfy the ravenous appetite for materials consumption in the developed and the developing world. Many Indigenous communities are using their creation myths and art and literature as a means of resistance. In this sense Indigenous creation myths have taken on new significance for our time. Their purpose is not so much to answer the question, Where do we come from? But rather to address, Where are we going? Adamson (and others in this volume) make similar observations. Indigenous stories are being effectively used as "seeing instruments" by ecocritics and activists "for making abstract, often intangible global patterns associated with climate change, accessible to a wider public" (Adamson 2013, 172).

The Role of Time in Long Environmentalism

Toward the end of the nineteenth century, Plenty Coups, the great chief of the Crow Nation, shared his sentiment with an outsider: "When the buffalo went away the hearts of my people fell to the ground; they could not lift them up again. After this nothing happened." Philosopher Jonathan Lear opens his rather slender book, *Radical Hope: Ethics in the Face of Cultural Devastation*, with Plenty Coups' statement and for the rest of the book makes philosophical inquiries into that last line—"After this nothing happened"—with regard to strategies for cultural survival (Lear 2006). The people of the Gwich'in Nation fear that oil development in the calving ground of the Porcupine River caribou herd on the coastal plain of the Arctic National Wildlife Refuge would destroy the herd, and subsequently, the Gwich'in culture. Through a poster that reads, "Will the caribou go the way of the buffalo? Or will you save our Arctic way of life?" the Gwich'in Nation explicitly connected the fate of the buffalo and the plains Indians with the possible fate of the caribou and the Gwich'in. Sarah James said, "We are the caribou people. Caribou are not just what we eat; they are who we are. They are in our stories and songs and the whole way we see the world. Caribou are our life. Without caribou we wouldn't exist" (James 2013, 262). Her statement expresses similar concerns to those of Plenty Coups. There is also a key common ground in their strategies for survival—collaboration. Plenty Coups collaborated with the U.S. government—an unlikely ally—for the survival of his people, even as their way of life was being destroyed and they had to accept a new way of life on the reservation. Lear calls this "Radical Hope." Similarly, the Gwich'in collaborate with conservation groups—traditionally unlikely allies—to help them fight for cultural survival. While Plenty Coups lamented the destruction of the way of life of the Crow people that he had witnessed, Sarah James, by contrast, is staking a claim on the future survival—"Without caribou we wouldn't exist"—of the Gwich'in way of life as they know it today.

Land conservation in the U.S. began with a rift, marked by deep injustices against the Native Americans, but more than a century and half later conservationists and the Indigenous peoples are beginning to come together to oppose destructive, large-scale corporate extractive projects and government policies that support them. In case of the Arctic National Wildlife Refuge campaign, the two opposed ideas—"deep ecology" and "environmentalism of the marginalized"—seem to co-exist to avert potential social-environmental destruction.[3] This has been possible because of long environmentalism, which in this case has lasted for more than six decades and is ongoing.

When a social-environmental engagement has lasted for more than six decades, it naturally becomes intergenerational. In a letter to U.S. Senator Daniel Akaka, young Gwich'in writer-activist Matthew Gilbert (2013, 484) poignantly illuminated the intergenerational attribute in the Arctic

Figure 3.3 Gwich'in caribou harvest, Arctic National Wildlife Refuge near Arctic Village, Alaska. January 2007. Source: Courtesy of Subhankar Banerjee.

National Wildlife Refuge campaign. Reminiscing about the 1988 gathering in Arctic Village, Gilbert writes, "As a kid, I remember the leaders with their traditional talking sticks on stage speaking passionately. Though at that young age, the topic was new and unfamiliar, I nonetheless saw the sincerity of the Gathering and respected it, even as a kid." He also highlights his own engagement in the campaign with these words: "I've been protesting against opening the Arctic National Wildlife Refuge to oil and gas drilling since I was nine years old. I was an innocent kid holding up two marker-colored banners made by two other kids, posing for a photographer. I ended up in *Time* magazine. From there, it never stopped" (480).

You might be wondering what role time plays in long environmentalism. Time enables the "marginal edge" and the "radical edge" to move from the periphery and toward the center, deflating the circle of power by challenging race and class induced injustices. The Gwich'in people were considered marginal (or even nonexistent) during the formation of the Arctic National Wildlife Range six decades ago, and the idea to allow subsistence activities inside wilderness would have been considered radical (or even preposterous) back then. And yet today, the Gwich'in are significant agents in the Arctic Refuge campaign and Indigenous subsistence activities inside the (Alaska) wilderness are considered essential. This is why the need to keep radical hope alive over a long period of time is so important and is an essential tenet of long environmentalism.

Slow Violence Requires Long Environmentalism: Point Hope and Beyond

The photograph, *After the Listening Session*, weaves two braids of long environmentalism—resistance to destruction as exemplified in the Arctic National Wildlife Refuge campaign and response to slow violence that I will now discuss briefly in a second case study. This case study is epitomized by renowned Iñupiaq cultural activist Caroline Cannon's testimony in support of a lawsuit against the U.S. Department of Interior, filed in 2009. The statement captures the ethos of the battles Indigenous peoples have had to fight all over the world: "We must fight and do all we can to preserve our way of life even if we feel like we have been fighting the same fight for the last *fifty years*" (Cannon 2013, 320; emphasis mine).

Cannon lives in Point Hope, a community of about eight hundred residents. Point Hope is situated along the Chukchi Sea coast in Arctic Alaska and is considered to be one of the oldest continuously inhabited settlements in North America. With the words "same fight," Cannon links two distinct and consecutive struggles, to "preserve" the "way of life" of the Iñupiat people against first, militarization, and then, industrial incursion.

In 1958, nuclear physicist Edward Teller, considered to be the father of the hydrogen bomb, went to Alaska to promote Project Chariot—to create a deep-water harbor at Cape Thompson, about thirty miles southeast of Point Hope, by detonating a string of nuclear bombs. Appropriately alarmed by the potentially devastating consequences of nuclear contamination on people and on biotic life, the Iñupiat people of Point Hope, with help from biologists Leslie Viereck and William Pruitt, geographer Don Foote, and a handful of conservationists, including Ginny Wood, launched a courageous and creative campaign to stop the project (O'Neill 1994). After a drawn-out and acrimonious fight, Project Chariot was shelved in 1963. The project created a deep scar in the psyche of the people of Point Hope. Cannon explains in her testimony:

> With Project Chariot, the federal government took advantage of us. They treated us like we were nonexistent people ... They were ready to relocate us and told us that the radiation wouldn't harm us. They took something away from us then. It was trust. We were emotionally damaged—feeling that we didn't count, that we were nothing.
>
> (Cannon 2013, 320)

Critics, including Ramachandra Guha, have charged that environmentalism in the U.S. rarely engaged with, or challenged, the nation's growing militarism. The campaign to stop Project Chariot should be considered an exemplary exception. It is also likely the first major grassroots environmental movement in the U.S. in which Indigenous people, conservationists, and scientists worked together to oppose social-environmental destruction through militarization.

Even though the bombs were not detonated, the U.S. government performed a nuclear experiment at Cape Thompson in 1962 and buried the waste. Thirty years later when clean up began, low levels of radioactivity were detected at a depth of two feet from the surface in the area where animals and people cross paths. In 2007, conservationist and author Peter Matthiessen and I visited Point Hope (Matthiessen 2007). During our conversation with the village council members (including Caroline Cannon), we learned that the community members believe that Project Chariot irradiated the animals and the people. "Many of our young people have died of cancer," Cannon writes. "My own daughter was diagnosed with leukemia in August of 2005, which is known to be linked to exposure to radiation" (Cannon 2013, 320). The environmental degradation and the consequent health impacts that the people of Point Hope have endured could be apprehended as *slow violence*. As postcolonial literary scholar Rob Nixon writes, slow violence occurs "gradually" and remains "out of sight" for the mainstream society (Nixon 2011). How do communities respond to slow violence? As this case study illustrates, slow violence, if it is to be addressed successfully, requires a long, concerted environmentalism.

The struggle against one environmental degradation from Project Chariot spilled over into another—this time, against oil and gas drilling in the Chukchi and Beaufort seas of Arctic Alaska, which is of great concern not only for residents of Point Hope but also for many other Iñupiat who live in various communities along the Arctic sea coasts in Alaska. The Iñupiat people, who value their traditional culture, fear that a blowout from drilling activities in the Arctic seas may ruin the millennia-old relationship they have built with the sea. They depend on the sea and its biotic life for nutritional, cultural, and spiritual needs, not unlike how the Gwich'in people depend on the caribou. In a cold, early November morning in 2001, I photographed an Iñupiat cemetery on Barter Island along the Beaufort Sea coast. The cemetery is marked by a pair of bowhead whale jawbones, which seems to honor both the family members who passed away and the whale that fed the community.

The first wave of oil development in the Beaufort and Chukchi seas began in the late 1970s and lasted through the early 1990s. The expensive hunt for oil in the Arctic seas, however, largely failed, as exploration did not lead to production, except in one case, and that too not in far offshore but a nearshore, anchored-to-the-ground facility (LeVine, VanTuyn, and Hughes 2014). As a consequence, the companies relinquished almost all of the leases they had purchased. The second wave of U.S. Arctic offshore oil and gas activities started when George W. Bush took office in 2001. Between 2003 and 2008 leases were sold on more than three million acres in the Beaufort and Chukchi seas. The lease sales generated substantial controversy and met with court challenges from the tribal Iñupiat and conservation organizations, for which I wrote supporting declarations. The plaintiffs won twice, first in the U.S. District Court for the District of Alaska in 2010, and then in the Ninth Circuit Court of Appeals in January 2014. But such victories were only short lived.

Figure 3.4 Iñupiat cemetery marked by bowhead whale jawbones, Barter Island, along the Beaufort Sea coast, Alaska. November 2001. Source: Courtesy of Subhankar Banerjee.

A *Deepwater Horizon*-like blowout in the Arctic seas would be devastating, worse than what happened in the Gulf of Mexico in 2010 (Banerjee 2013). For a variety of reasons, drilling for oil and gas in the Arctic Ocean is likely the most dangerous industrial project on Earth, as there is no proven technology to clean up an oil spill from underneath sea ice in one of the harshest environments on the planet. Even during the open-water season, there is frequent fog and severe storms, as well as large ice floes—all of which would make effective clean-up very difficult, if not impossible. The Arctic is also warming at a rate of about two to four times the global average, which has significant impacts on the Arctic ecology and the Indigenous communities. Oil and gas drilling in the Arctic seas would only add further stress and devastation. For example, it would exacerbate Arctic warming from various pollutants that industrial operations would generate (Banerjee 2015a). Moreover, extraction of fossil fuel resources from the Arctic is "incommensurate with efforts to limit average global warming to 2°C" above the pre-industrial level (McGlade and Ekins 2015, 187–190). Despite all of these concerns, as well as the continued lack of comprehensive scientific understanding of the marine ecology in the Beaufort and Chukchi seas, the Obama administration granted Shell Oil the necessary permits to drill in the Chukchi Sea during the open-water season in 2015.

Overall, two interrelated attributes have emerged from the push to industrialize America's Arctic Ocean, which are made visible by long environmentalism: the *inevitability* and the *rush*. The "inevitability" is a mindset that drilling in U.S. Arctic waters is inevitable, as opposed to protecting those seas as international ecological treasures replete with tens of thousands of whales, thousands of polar bears, hundreds of thousands of walruses and seals, millions of birds, and innumerable fish, not to mention all the tiny sub-sea life that make up the food chain. The inevitability I speak of has been exhibited repeatedly in Arctic science research sponsored by the U.S. government, including in an important booklet, *Arctic Matters: The Global Connection to Changes in the Arctic* (Polar Research Board and National Research Council 2015). Also consider this as another example: In 2015, the U.S. Fulbright Program launched the Fulbright Arctic Initiative (Council for International Exchange of Scholars n.d.). Of the four research areas, the first on the list is "Energy" and the first question asked is this: "How will oil, gas, and other natural resources be developed in the Arctic?" The keyword "ecology" is missing from the Fulbright Arctic Initiative research goals. I mention the omission because environmental humanists are now paying particular attention to ecology and other critical "Keywords for Environmental Studies" (Adamson, Pellow, and Gleason 2016).

The "rush," on the other hand, is a practice of both the fossil fuel industry and the government (Banerjee, 2015b). These examples illustrate why there is a need for research in Arctic humanities, in addition to Arctic science, so that a more just future for the Arctic can be envisioned that is not merely informed by an ideology of exploitation.

Long environmentalism here, which began with a grassroots movement against Project Chariot, has evolved over six decades incorporating legal challenges and exposing continued injustices in the U.S. government's attitude toward Indigenous peoples. Cannon writes that "the government and industry continue to ignore our concerns and run roughshod over our community." But she also writes: "We have a right to life, to physical integrity, to security, and the right to enjoy the benefits of our culture. For this, we will fight" (Cannon 2013, 327).

Figure 3.5 Prayer after an Iñupiat whale hunt to thank the Creator and the whale for offering food for the community. Barter Island, along the Beaufort Sea coast, Alaska, September 2001. Source: Courtesy of Subhankar Banerjee.

After nine years of trying to pry open the Arctic seas for oil, spending more than $7 billion, and receiving all the necessary permits for exploratory drilling, Shell announced on September 28, 2015, after a brief season of exploration in the Chukchi Sea, that the company has abandoned its Arctic Alaska offshore program, citing "disappointing exploratory results, high operating costs, and strict U.S. environmental regulations" (Neslen 2015). A few weeks later, the Obama administration announced that it was canceling future lease sales in those seas. It would be safe to say then that the second wave of Arctic offshore development has come to an end. This is very significant news for climate change mitigation and a relief for the Iñupiat people, who value their traditional culture.

Coda

A study of long environmentalism is a work in progress. In summary, then, I return to the four functions mentioned in the introduction that long environmentalism performs, the first three of which are addressed by the two braids of long environmentalism embedded in *After the Listening Session* photograph: it illuminates past injustices, as we have seen in both case studies; it highlights the significance of resistance movements to avert potential social-environmental violence (fast and/or slow), as we have seen in the Arctic National Wildlife Refuge campaign; and it shows how communities respond to slow violence, as we have seen in the Iñupiat people's struggles against destruction. The fourth function, which points toward social-ecological renewal after devastation, while not addressed in the two case studies discussed here, can be seen in other instances of long environmentalism, as in the case of the current restoration of the Elwha River in the Olympic National Park in Washington State following the recent destruction of two dams that were built a century ago and which destroyed the epic salmon runs by preventing the fish from reaching their ancestral spawning grounds and consequently the culture and nutritional needs of the Lower Elwha Klallam people. In all of these cases, both collaboration among unlikely allies made possible through sincere listening proved essential, and radical hope was kept alive over decades, proving that those two attributes are essential for long environmentalism. In the Anthropocene, varieties of environmental violence are here to stay, will likely get worse, and many new ones will arrive. The inevitable social response, it seems to me, is long environmentalism.

Notes

1. This essay grew out of several conference talks the author gave on the theme of "long environmentalism": a keynote lecture at the "PostNatural" conference of the twenty-seventh Annual Meeting of the Society for Literature, Science and the Arts at the University of Notre Dame in 2013; and three talks in 2015: the inaugural lecture for a yearlong series, "Environmental Humanities,"

at the University of Texas–Austin, an Andrew W. Mellon Foundation Visiting Artist lecture at the University of New Mexico–Albuquerque, and a lecture at the "Conflict Shorelines: History, Politics, and Climate Change" conference at Princeton University.

2. A selection of my Arctic photographs was included in the *Rights of Nature* exhibition. I also participated in both conferences and pointed out that the contemporary invocation of rights of nature is distinctly different from what Nash wrote about, and the difference is this: one takes into account the survival concerns of the poor and marginalized human communities, while the other does not. Both, however, are concerned about the survival of nonhuman biotic life.
3. From my years of working with Indigenous communities in the Arctic, I have learned that even when a community is struggling financially, as many rural communities in Alaska are, the people take offense when someone refers to them as "poor." They consider themselves rich, because of the still-abundant biotic life on the land, in the rivers, and in the oceans on which they depend on for nutritional, as well as cultural and spiritual, needs. To be able to accommodate, without offending, communities from the circumpolar North in the discourse of the "environmentalism of the poor," I have chosen to use the term "marginalized" instead.

References

Adamson, Joni. 2001. *American Indian Literature, Environmental Justice, and Ecocriticism: The Middle Place*. Tucson: The University of Arizona Press.

———. 2013. "Environmental Justice, Cosmopolitics, and Climate Change." In *The Cambridge Companion to Literature and the Environment*, ed. Louise Westling, 169–83. Cambridge, UK: Cambridge University Press.

Adamson, Joni, Mei Mei Evans, and Rachel Stein, eds. 2002. *The Environmental Justice Reader: Politics, Poetics, and Pedagogy*. Tucson: University of Arizona Press.

Adamson, Joni, David Naguib Pellow, and William A. Gleason. 2016. *Keywords for Environmental Studies*. New York: New York University Press.

Banerjee, Subhankar. 2003. *Arctic National Wildlife Refuge: Seasons of Life and Land*. Seattle, WA: The Mountaineers Books.

———.2013. "BPing the Arctic?" In *Arctic Voices: Resistance at the Tipping Point*, ed. Subhankar Banerjee, 66–90. New York: Seven Stories Press.

———. 2015a. "In the Warming Arctic Seas." *World Policy Journal* 32 (2): 18–27.

———. 2015b. "To Drill or Not to Drill, That Is the Question: The Obama Administration, Shell, and the Fate of the Arctic Ocean." *Tomdispatch* March 3. Accessed January 15, 2016. http://www.tomdispatch.com/post/175963/tomgram%3A_subhankar_banerjee%2C_arctic_nightmares.

Bloom, Lisa. 1993. *Gender on Ice: American Ideologies of Polar Expeditions*. Minneapolis: University of Minnesota Press.

Cannon, Caroline. 2013. "Testimony in Support of a Chukchi Lease Sale 193 Lawsuit, Case No. 1:08-cv-00004-RPB (2009)." In *Arctic Voices: Resistance at the Tipping Point*, ed. Subhankar Banerjee, 319–28. New York: Seven Stories Press.

Ceballos, Gerardo, Paul R. Ehrlich, Anthony D. Barnosky, Andrés García, Robert M. Pringle, and Todd M. Palmer. 2015. "Accelerated Modern Human–Induced Species Losses: Entering the Sixth Mass Extinction." *Science Advances* 1, no. 5 (June 19). Accessed January 15, 2016. http://advances.sciencemag.org/content/1/5/e1400253.

Council for International Exchange of Scholars. n.d. "Fulbright Arctic Initiative." *Fulbright Scholars Program* (website). Accessed 5 May 2016. http://www.cies.org/program/fulbright-arctic-initiative.

Demos, T. J. 2015. "Rights of Nature: The Art and Politics of Earth Jurisprudence." (essay). Associated with the exihibit, *Rights of Nature: Art and Ecology in the Americas*. In *Nottingham Contemporary* (website). Accessed January 15, 2016. http://www.nottinghamcontemporary.org/sites/default/files/Rights%20of%20Nature%20The%20Art%20and%20Politics%20of%20Earth%20Jurisprudence.pdf.

Dunaway, Finis. 2009. "Reframing the Last Frontier: Subhankar Banerjee and the Visual Politics of the Arctic National Wildlife Refuge." In *A Keener Perception: Ecocritical Studies in American Art History*, ed. Alan Braddock, and Christophe Irmscher, 254–74. Tuscaloosa: University of Alabama Press.

Esbjörn-Hargens, Sean, and Michael E. Zimmerman. 2009. *Integral Ecology: Uniting Multiple Perspectives on the Natural World*. Boston, MA: Shambhala Publications.

Francis. 2015. *Laudato Si'*. Encyclical letter on care of our common home. Vatican encyclicals. Libreria Editrice Vaticana. Accessed 5 May 2016. http://w2.vatican.va/content/francesco/en/encyclicals/documents/papa-francesco_20150524_enciclica-laudato-si.html.

Gemmill, Faith. 2001. "Caribou Story." In *Arctic Refuge: A Circle of Testimony*, ed. Hank Lentfer, and Carolyn Servid, 48–51. Minneapolis, MN: Milkweed Editions.

Gilbert, Matthew. 2013. "Letter to Senator Daniel Akaka about Arctic National Wildlife Refuge (2005)." In *Arctic Voices: Resistance at the Tipping Point*, ed. Subhankar Banerjee, 480–85. New York: Seven Stories Press.

Guha, Ramachandra. 1989. "Radical American Environmentalism and Wilderness Preservation: A Third World Critique." *Environmental Ethics* 11 (1): 71–83.

Guha, Ramachandra, and Juan [Joan] Martinez-Alier. 1997. *Varieties of Environmentalism: Essays North and South*. London: Earthscan.

———. 2000. *Environmentalism: A Global History*. New York: Longman.

"Gwich'in Elders Statement." 1988. *Oil on Ice*. Accessed 5 May 2016. http://www.oilonice.org/explore/eldersstatement.php.

Jacoby, Karl. 2001. *Crimes against Nature: Squatters, Poachers, Thieves, and the Hidden History of American Conservation*. Berkeley: University of California Press.

James, Sarah. 2013. "We Are the Ones Who Have Everything to Lose." In *Arctic Voices: Resistance at the Tipping Point*, ed. Subhankar Banerjee, 260–66. New York: Seven Stories Press.

Lear, Jonathan. 2006. *Radical Hope: Ethics in the Face of Cultural Devastation*. Cambridge, MA: Harvard University Press.

LeVine, Michael, Peter Van Tuyn, and Layla Hughes. 2014. "Oil and Gas in America's Arctic Ocean: Past Problems Counsel Precaution." *Seattle University Law Review* 27 (4): 1271–1370.

Martinez-Alier, Joan. 2003. *The Environmentalism of the Poor: A Study of Ecological Conflicts and Valuation*. Cheltenham, UK: Edward Elgar Publishing.

Matthiessen, Peter. 2007. "Alaska: Big Oil and the Iñupiat-Americans." *New York Review of Books*, 22 November. http://www.nybooks.com/articles/2007/11/22/alaska-big-oil-and-the-iñupiat-americans/.

McGlade, Christophe, and Paul Ekins. 2015. "The Geographical Distribution of Fossil Fuels Unused When Limiting Global Warming to 2° C." *Nature* 517 (January 8): 187–90.

Nash, Roderick F. 1989. *The Rights of Nature: A History of Environmental Ethics*. Madison: University of Wisconsin Press.

Nixon, Rob. 2011. *Slow Violence and the Environmentalism of the Poor*. Cambridge, MA: Harvard University Press.

Neslen Arthur. 2015. "Shell Has Frozen Its Arctic Oil Drilling—But It's Still Hungry for Fossil Fuels." *The Guardian*, September 28. Accessed January 15, 2016. http://www.theguardian.com/environment/2015/sep/28/shell-has-frozen-its-arctic-oil-drilling-but-the-fight-isnt-over.

O'Neill, Dan. 1994. *The Firecracker Boys*. New York: St. Martin's Press.

Polar Research Board, Division on Earth and Life Studies, and the National Research Council. 2015. *Arctic Matters: The Global Connection to Changes in the Arctic*. Washington, DC: National Academies Press.

U.S. Fish and Wildlife Service. 1960. *Public Land Order 2214*, by Fred A. Seaton. F.R. Doc. 60–11510. 6 December. http://www.fws.gov/refuge/arctic/plo2214.html.

Wallis, Velma. 2013. "Past and Present, Culture in Progress." In *Arctic Voices: Resistance at the Tipping Point*, ed. Subhankar Banerjee, 486–98. New York: Seven Stories Press.

Williams, Maria. 2009. "A Brief History of Native Solidarity." In *The Alaska Native Reader: History, Culture, Politics*, ed. Maria Williams, 202–16. Durham, NC: Duke University Press.

Zimmerman, Michael. 2015. "Integral Ecology in the Papal Encyclical, On Care for Our Common Home (*Laudato Si'*): A Critical Appreciation (Blog post). *Institute for Cultural Evolution*. 7 July. http://www.culturalevolution.org/blog/how-integral-is-the-recent-papal-encyclical-on-care-for-our-common-home/#.Vph8jbz9nUo.

4 Grounded in Spiritual Geography

Restoring *Naabaahii* in *Enemy Slayer, A Navajo Oratorio*

Laura Tohe

Musicians and composers brighten and humanize the world. While I am not a composer, music has been my lifelong passion. Before I could appreciate some forms of music, like opera, I had to mature into it. Throughout my literary life I have written poetry, essays, and short fiction, and published my work locally and internationally. Though I didn't like or appreciate opera when I was younger, ironically, I came to write the oratorio, *Enemy Slayer, A Navajo Oratorio*, a commissioned work for the Phoenix Symphony (PS) under the direction of Michael Christie, who was the Virginia G. Piper Music Director at the time. Mark Grey, a composer, was commissioned to write a musical composition as part of a new work for the PS's sixtieth anniversary celebration. Preliminary discussions between Michael and Mark, in which I was not involved, raised the possibility of creating a work based on a creation story from one of the local Indigenous communities, perhaps, the story "Naayéé neizghání" ("Monster Slayer"), from the Navajo mythic stories. The Navajo are also known as the *Diné*, which translates into English as "The People." We are also known as the Navajo. I will use these terms interchangeably. My involvement with the project began later with Mark's search for a Diné poet to write the libretto, the lyrics for the performance. We corresponded, and after looking over his website and speaking with him, I accepted his offer to write the libretto without knowing what an oratorio and libretto were. I looked the terms up later. While the oratorio is a form of opera, it is not as full-blown, with elaborate stage sets and a large cast. An oratorio usually tells a biblical creation story, with a single or a few soloists, an orchestra, and a chorus. Handel's *Messiah* and Haydn's *The Creation* are among the most well-known oratorios that were commissioned in Europe in the eighteenth century.

A few months later we began the collaboration on *Enemy Slayer, A Navajo Oratorio* that would take us over a year and a half to complete for the world premiere in 2008 at Phoenix Symphony Hall, where it reached new audiences in the hour-long performance. *Enemy Slayer, A Navajo Oratorio* was recorded on the Naxos classical music label and was also performed at the Colorado Music Festival and at the University of Utah in Salt Lake City. After the Phoenix world premiere, guests were invited to post comments on the PS's web page. One post said the production was staged like a Navajo ceremonial assemblage—with Michael Christie as the medicine man who conducted the ceremony and Scott Hendricks as the patient, while the female and male chorus members and the orchestra represented the extended family. I deeply appreciate the vision of Michael Christie, who commissioned this new kind of work from the Arizona Indigenous peoples, and the generous artistry and sensitivity of Mark Grey, who composed the music. Scott Hendricks's powerful baritone voice and insight brought forth Seeker's pain and rejuvenation, while Deborah O'Grady's photographic design contributed a visual narrative to the oratorio. The PS's 100-member chorus and full symphonic orchestra gave voice and presence to the rich oral tradition of the Diné within a symphonic setting, an achievement not many musical organizations have accomplished. *Enemy Slayer, A Navajo Oratorio* premiered as a groundbreaking hybrid of the European oratorio and Diné oral tradition.

Figure 4.1 Rehearsal for *Enemy Slayer, A Navajo Oratorio,* Phoenix Symphony Hall, February 5, 2008. Source: Courtesy of the author.

Though the *Enemy Slayer* performances were a success, the initial conception of the project as a creation story raised red flags among traditional Diné elders living in Phoenix, including myself. The project sought to bring together the Navajo community, the PS's subscription audience, and the greater Phoenix community. The original synopsis of the oratorio was intended as a Navajo creation story of the Hero Twins, as oratorios are often about creation stories. The Hero Twins, epic and powerful, are the sons of White Shell Woman and the Sun. In Navajo oral traditions, it is said that in the previous Fourth World, the Twins took a journey to the Sun to ask for weapons to make the world safe from the monsters that were killing the people. Nayenezgani (correctly spelled as Naayéé neizghání), or Monster Slayer, is the name for one of the twins. However, writing a creation story about about Monster Slayer raised a red flag, and I knew it would concern the traditional Navajo people as well.

Navajo creation stories and sacred ceremonies, such as The Enemy Way ceremony, which the Hero Twins are associated with, are closely guarded to protect their integrity and sacred healing power. The Enemy Way ceremony is a complex healing ceremony and one of its uses is to treat post-traumatic stress disorder (PTSD) so that the veteran may be restored to a state of well-being, spiritually, and psychologically, and to a healthful return to a productive and peaceful life among family and community. This ceremony brings together the healer (the medicine man), the extended family, and the community to actively participate in the restoration of their loved one. With the healer's spiritual guidance, ceremonial prayers and songs, healing thoughts, and the collective spirit of the clan, the patient's extended family actively participates in the ceremony's healing power to restore mind, body, and spirit. The Laguna people, who are neighbors to the Diné, also recognize the need for this kind of ceremony. In Leslie Marmon Silko's *Ceremony* (1977) the character Tayo is placed in the hands of Betonie, the Navajo medicine man. While the Enemy Way ceremony is specific to the Navajos, Tayo's immersion into healing traditions enables him to complete a broken cycle before he can return to his community free from PTSD.

The PS's mission statement to collaborate and provide outreach activities with local schools and cultural groups, including the Navajo Nation, rooted this work in the community, and people began discussing it. Word came back to us that a group of Diné elders living in Phoenix were raising issues about using a Diné creation story as a public performance. For this reason, before the first musical note and words were written, Mark and I met with them so that they could express their concerns. They were right to question the use of Diné sacred stories, let alone a creation story as public entertainment. The elders who raised issues spoke up to protect the cultural sovereignty of the Diné. I have seen and heard about theatrical performances that used Indigenous oral traditions that were misrepresented or misappropriated. While the creators of public performances by non-Native peoples may have good intentions, the results often reinforce negative stereotypes, distort, or caricature the images and cultures of Native peoples, such that the end results create disrespect for Native peoples' stories and their oral traditions. For example, several years ago I collaborated on writing a tri-cultural children's play intended as an affirmation of cultural ancestry that the young audience could learn from. When it was produced on stage, I was horrified that the Navajo characters were dressed in feathered headdresses, even though my tribal nation doesn't wear elaborate headdresses like other tribes. Instead of learning that tribal nations have unique cultures and languages, including modes of dress, the young audience was presented with the stereotype of the Indian. Others, including myself, who have had similar experiences took from this that the integrity of our work must be protected from the norms of the dominant culture and from stereotypical representations. As artists, we must insist upon this so that Native works are true representations of our diverse cultures and identities.

In a private meeting with my grandfather, who was at the oratorio meeting with the elders, I asked what he thought of this project. I wanted his honest concerns about writing the libretto. We met again after he consulted with his mother. Her comments raised the potential that the oratorio would create controversy if it took shape as a creation story of the Hero Twins, as originally planned. She was concerned that the audience, who would mostly be non-Diné, would not be able to understand the contexts surrounding Navajo creation stories. As Scott Momaday writes in his essay "The Man

Made of Words," the history of colonialism has resulted in the "theft of the sacred" and the loss of cultural objects and sacred traditions (Momaday 2001, 76). The protection and preservation of sacred ceremonies and stories are part of tribal national sovereignty and cultural identity and must be upheld and preserved for future generations. Further, Momaday says that an Indian is who *he imagines* himself to be (2001, 87). By extension we can see how the Diné *imagine* themselves as a people who believe in the power of language to heal and restore.

It is also worth considering how Indigenous people define and use the word "sacred" to apply, not only to spiritual beliefs, but also to the secular. When Diné people use this word, it can be linked with almost everything—"sacred water," "sacred buffalo," or "the four sacred mountains"—while in the Western world this term has lost much of its consecrated meaning, except within the context of religion and what is held as holy. Diné ceremonies and stories are part of cultural identity and they define relationships to the land and sky that chart ceremonial activities based on celestial movements and the times of the day. For these reasons, there was the possibility that a non-Indian audience, who might know little or nothing about the Indigenous peoples of North America, let alone Diné culture and stories, would not understand the complex meanings in creation stories that are considered sacred. They could become misappropriated and misinterpreted, my grandfather cautioned. We didn't speak of other ways to write this oratorio, but he affirmed my own concerns. The traditional way of advising is to lay out the possible scenarios, as my grandfather did. I was free to choose my path and, consequently, I would be responsible for what I created with words. Telling people what choice to make not only takes away responsibility but also erodes the assumption that they have the intelligence and maturity to make their own decisions.

Mark and I discussed various storylines that would appeal to a multicultural audience and one that would not be offensive to the Diné. I had recently made a trip home to my rez in northern Arizona. As I drove Interstate 40, which runs parallel to the railroad tracks near Winslow, Arizona, I noticed miles and miles of freight moving west toward California. The flatbed cars were loaded with military equipment—tanks, Humvees, and artillery. It was 2006, and the Middle Eastern war dominated the news media. It was taking many lives, mostly the young, and vast numbers of tax dollars. President George W. Bush refused to allow television news to film the caskets arriving home on U.S. soil. Instead viewers saw only the names of the fallen soldiers. I felt strongly that the libretto should say something about the war and the tremendous costs to our families and our nation. I thought a great deal about the project and how we could create a hybrid work of Indigenous storytelling, not only to speak of the war but also to give visibility to the Navajo oral tradition without creating the kind of controversy that erodes its integrity and sacredness. We dispensed with a creation story and decided on a story about a veteran who returns from the Middle Eastern war with PTSD and is grounded in the Diné worldview of the four directions, symbolized by the four sacred mountains that sit at each cardinal direction of the Diné homeland. From this scaffold we set out to create the oratorio, *Enemy Slayer, A Navajo Oratorio*.

Four major mountains mark the original Diné homeland. Before the arrival of the Europeans, the geographical space within the boundaries of these mountains was named *Diné Bikéyah,* The People's Homeland, by the Navajo people. Blanca Peak, or *Sis Naajiní,* located in southern Colorado sits as the eastern mountain. Mount Taylor is called *Tsoodził* and sits near Grants, New Mexico, as the southern mountain. Near Flagstaff, Arizona, sits the San Francisco Peak or *Dook'o'oosłííd,* and it is the western mountain. The La Plata Mountains, also in southern Colorado, are the northern mountains and are named *Dibé Nitsaa*. Navajo stories of long ago say that the space within the four sacred mountains is where the Diné found agreement with the land and where they would live and sustain themselves. Prior to the arrival of the Spanish and EuroAmericans, the Diné moved freely within and beyond this space until they were forced onto a reservation with the signing of the Navajo Treaty of 1868. Most of the former ancestral land base of the Diné people was taken over through colonial appropriation. Nevertheless, the Diné still recognize the land within the four sacred mountains as *Diné Bikéyah,* even though none of the mountains sit within the boundaries of the federally drawn Navajo reservation that is now in Arizona, New Mexico, and Utah.

The philosophy embedded within the four directions and four mountains form part of Diné epistemology and their cosmovision of the earth, which includes the sky. As such, the mountains need to be protected because they maintain stability and order throughout the life cycle of all living things, animate and inanimate. The protagonist, Seeker, enters the story in the Introduction at Blanca Peak and travels metaphorically in a clockwise direction to each of the mountains until he reaches the north mountain. His story unfolds in four parts as symbolized by each of the mountains and completes a cycle. Diné refer to these mountains as the Four Sacred Mountains and consider them to be sentinels, protectors of the homeland, and they represent stability. More so, these mountains are imbued with spirit, intelligence, sacredness, and aesthetic and emotional sustenance. Mountains inspire our human sense of the majestic by their physical beauty and grandeur. George Blueeyes, Diné elder, says in his poetic expression about the mountains, "Our Navajo laws are represented by the/Sacred Mountains … They were placed here for us … We think of them as home" (Evers 1980, 2). The Diné sense of place is defined by their relationship to the mountains and the earth as spiritual and historical within the geographic space of the four mountains, and from this relationship they have given place names to mountains, rivers, valleys, canyons, rock formations, and other natural features found on The People's homeland, *Diné Bikeyah*. Since time immemorial the construct of Diné identity is closely associated with this land. In this space Diné would live their lives and establish a reciprocal relationship with Nahasdzáán, Mother Earth. The People gave meaning to this geographic place and their interaction with it was thought of as *home*, the place where mothers bury their children's umbilical stems near the home so that their children's roots will remain connected to the earth and will act as a kind of map to *home*.

My family and elders told me epic stories that took place within this space, stories of emergence from previous worlds and battles with monsters that killed humans until the twin brothers, Monster Slayer and Child Born for Water, made a remarkable trip on a rainbow to seek their father, the Sun, and to ask for his help to secure weapons that they could take back to earth to destroy the monsters. On one of our many car trips to Albuquerque, New Mexico, from our home, my mother told of how one of these giants was slain near what is now Grants, New Mexico. The solidified lava flow that we were driving on below the interstate was proof of the giant's blood. I always thought it extraordinary how my mother's stories and the stories the Diné tell link story with place and how *Diné Bikeyah* is essentially a map of stories from time immemorial to the present day. The Twins killed most of the monsters and made the world safe. Soon afterwards the brothers suffered with depression and the symptoms of what, today, is called PTSD. At a loss for how to help the Twins, The People prayed for help from the Holy People and were given the Enemy Way ceremony through divine means to restore the Twins. It is still used today to heal veterans from the U.S. wars they have fought, including the Middle Eastern wars. The story of the Hero Twins provides the context for the oratorio. The story is set in modern times and tells the story of a veteran grounded in Navajo culture who returns from a war in the Middle East with PTSD and must find *Hózhôjí*, or spiritual balance. If he cannot find this balance, he may ultimately risk ending his own life and devastate his family and kinship communities and thereby violate the Navajo value to live a long natural life. By extension, his death would disturb the sacred relationship between the earth and the human spirit and the philosophy inherent in the mountains and the four directions that give guidance, support, and stability to the Diné. As George Blueeyes elucidates in his untitled poem about mountains, "… they have regulated our lives/since the beginning" (Evers 1980, 2).

Grounded within the boundaries of the four mountains that sit at each cardinal direction of the Diné/Navajo homeland, Seeker is aligned with Monster Slayer, a mythic hero in the Diné oral tradition, who returns home from war hailed as a hero but also returns with the monsters of war. Through the mythic and healing traditions of the Diné, the story of Seeker's return home with PTSD unfolds and, eventually, marks his journey toward restoration of spiritual health and balance, while an anti-war response underlies it. The mountains provide the framework for the story told in five parts with a return to the east at the end of the oratorio. Each cardinal point symbolizes a stage in human life, and, in this case, the development of the storyline.

Prologue

A prologue generally sets a story and gives contextual introductory information or background. *Enemy Slayer, A Navajo Oratorio* opens with the chorus singing a ten-minute Prologue that further establishes the geographic and cosmic context for Seeker, the protagonist, who returns to *Diné Bikeyah*. "[Seeker returns] from across the big water / Traveling lightly on a rainbow." The chorus members function like in a Greek chorus as they comment on Seeker's emotional and spiritual states and, at times, must lift him from the state of anomie to provide him with moral and spiritual guidance, especially as he further descends into spiritual rootlessness. They offer him his true path in life by blending tradition with compassion.

To create a visual sense of the mountains and the Southwest landscape where the story takes place, Mark invited landscape photographer Deborah O'Grady, who has collaborated with other composers to create musical and visual meditations on landscape as public performance. Her striking landscape photography brought a broader visual design to the oratorio that shaped it as a multi-dimensional piece. Deborah had traveled to each of the four mountains to photograph them for the performance. She also accessed images of wounded soldiers from the public domain. At key points in the performance, Deborah projected her images as the backdrop of the stage of Symphony Hall, which heightened the mood and intensity of the oratorio. Serendipitously, she happened to fly over Shiprock, New Mexico, in a commercial flight and took many of the opening images for the Prologue, which were used to portray Seeker's homecoming. Deborah's images brought the audience closer to Seeker's story with her photographs of Seeker's homeland, while the photographs of the wounded in the Iraq War resonated with Seeker's pain.

The chorus also represents the voices of the mountains that establish stability and serve as markers for Seeker's journey calling him into his place between the earth and sky within the phases of a day. Placed within this context, Seeker, the warrior, is also aligned with Navajo astronomy as the constellation *Átsé ets'ózí* (First Slender One, also known as Orion) envisions a young warrior holding a bow with a quiver of arrows strapped on his back, ready to strike. Seeker's life expectation is laid out among the pattern of stars that pronounce him *Naabaahii*, the ideal warrior who protects his family and home. "Implied in the name *Átsé ets'ózí* is the cyclical emergence of the constellation in a vast, cosmic process" (Maryboy and Begay 2010, 45). The chorus sings:

> Earth circling, stars revolving
> He steps into his place
> A warrior for his nation
> A warrior for the future

Instead of twin hero brothers, we chose one protagonist, Seeker, who seeks a way *home* from war. Since I had never written a libretto, or the lyrics, I was unsure of how to write them, even though I have written and published books of poetry and came to rely on that process as a starting point. My only instruction from Mark was to "write lines this long," and he held up his fingers to show three inches. In the Prologue I wanted to ground Seeker's identity within *Diné Bikeyah* as he returns to his homeland (where his umbilical cord is buried) and to his people, who believe in the spiritual connections to the four stages of a day and a human life. The Diné worldview is largely based on the number four: the four directions, the four sacred mountains, the four stages of life, the four phases of the day, and so forth. From this starting point Seeker is introduced as *Naabaahii*, the warrior/veteran returning home to a heartfelt welcome, wanting to reestablish himself there, with his family. "He calls forth, *shik'éí, shidiné'é*, meaning "my relatives, my people." I tried to keep translation issues at a minimum and used Navajo words that would be accessible to the singers. To say "my relatives, my people" is to include clan membership, which is the extended family, past, present, and future, including relatives one has never met. My mother used to point out a Navajo woman or man who was walking along the street in the local town where we shopped and would say, "there goes your grandmother" or "that man is your grandfather." I had never met that person, but I was to think of them as my relative, as we have relatives young and old who are considered grandmother, grandfather, aunt, and brother because of the clan membership. In a broader sense, Seeker is reestablishing himself with his relatives and with his homeland in the Prologue. He must use kinship terms to address his relatives and in doing so show familial bonds and responsible relationship, or as one Diné elder once said, "show a touch of love." To make Navajo pronunciation accessible for the chorus and the lead singer, I pre-recorded all the Navajo words and rehearsed the parts with them. Scott Hendricks, a much-in-demand and internationally known baritone singer and non-Native, was commissioned by the PS for the leading part. He had already been commissioned by the PS before I came on board. I recorded all the Navajo words and made copies for their rehearsals. Any trepidation on my part about the pronunciations dissolved. These were artists attuned to singing the nuances of sound and tone in many languages.

I have immediate family who enlisted in two wars; my father at age 16 was in the Marines during WWII as a Navajo Code Talker and two brothers who enlisted during the Vietnam era. My father rarely spoke of his service. "War is nothing to play at," he cautioned my brothers and me. In Peking (now Beijing) he recuperated in a hospital after WWII ended and was awarded a Purple Heart. What his injury was, I never knew. Of my two brothers who enlisted, the older was with the Army's Special Forces. He parachuted into the jungle where Agent Orange was sprayed to kill the foliage. While there he suffered a transportation accident that brought on seizures that would suddenly burst upon him throughout his short life. Like many veterans he became self-destructive and used alcohol to fight the war he brought home. I never witnessed my brother's terrible seizures that frightened my mother and grandmother. I can only imagine the nightmares that exploded in his sleep and awoke my mother. My grandmother once asked aloud, "I wonder what happened to him over there?" When he returned he bought a Dodge Charger and recklessly drove at high speeds.

As I wrote the libretto, I thought a lot of veterans like my brother, who passed at forty years of age and left a family. I thought of how war is never suffered alone by the veteran but also by families and nations who send their sons, daughters, brothers, fathers, and sisters into harm's way. Though veterans may return damaged by war, the ones left behind at home are also profoundly affected. Sometimes we are not able articulate those changes, but we feel them and we feel the wounds left by the fear of losing our loved ones. When my brother returned, I saw his difficulty in making his way back to his family and civilian life. I don't think my family realized the extent of his PTSD. Like other families, we probably thought it would eventually pass. But it didn't. He eventually entered a Veterans rehabilitation center for alcohol treatment and PTSD. Toward the end of his short life, his health and spirit were severely weakened. Upon visiting the Vietnam memorial in Washington, DC, I felt his name should have been etched on the monument, along with the others who returned with PTSD and were not memorialized. I stood with my ten-year-old son who asked why there was the list of names. After I explained he said, "even one name would have been too many."

Unlike the previous generation of Vietnam veterans who returned home without praise and public celebration, veterans of the Middle Eastern wars are often called heroes and given much public gratitude. Seeker, like his returning comrades, returns home welcomed as a hero, but the oratorio clearly suggests that he also brings the dark *monsters of war* that will hold him in their grip as his story progresses.

Cardinal Point One

Figure 4.2 *Blanca Peak*, 2007, Digital Image. Source: Courtesy of Deborah O'Grady, http://www.DeborahOGrady.com.

Blanca Peak, Colorado

East / Spring / Birth

Cardinal Point One begins in the east direction, which symbolizes birth, sunrise, and new beginnings. This is where Seeker's story begins when he returns home. Navajo people believe that a Navajo child is not only born into family and community memberships but also becomes a member of his mother's clan and is *born for* his father's clan. In this way, Navajo identity is constructed. Seeker establishes these facets of his identity when he sings, "I am called Seeker / Child of my mother's blood flow / Born for my father's people." Earlier the chorus broadened his identity when they placed him within the context of light and darkness as the earth revolves around the sun.

> Seeker, five-fingered being,
> Child of dawn
> Child of daylight
> Child of evening twilight
> Child of darkness

Seeker recalls his purpose in going to war as the *Naabaahii* whose traditional responsibilities are to protect home and nation, to not hesitate to take the right action, to have compassionate leadership, and to kill enemies. Seeker sees himself and his cousin brother (Diné term for a maternal cousin), who enlisted with him and, in a larger sense, symbolizes the tight brotherhood that soldiers form with their comrades. "Warriors for our nation / … We signed away our lives" but the cost of war extracts a price too. He mourns his brother who did not return with him.

> Brother, I miss you tonight
> In my mind I see the mound of earth
> and the plastic flowers baked by the sun that cover you now
> Black jet, black jet, black jet

Memories of his cousin brother's death, symbolized by black jet (black onyx), associated with the north mountain and the color for old age and death, begin to disturb him. He must mourn his brother's death, give it the needed time before moving on, and henceforth place his brother in the past tense whenever he speaks of him. The music slows to reflect Seeker's inner state of mind. To mourn excessively would be unhealthy for Seeker. The ancestral voices of the chorus recognize Seeker as their child and of the homeland. They offer him their blessings: *Bín' dii naabaahii bih'yázhí bik'i shándíín* ("Let sunlight shine on warrior child") and *Bín'dii nahasdzáán ãichíí' biyázhí nilîgo bótà* ("Let the red earth hold him as her child.").

Cardinal Point Two

Figure 4.3 *Mount Taylor*, 2007. Source: Courtesy of Deborah O'Grady, http://www.DeborahOGrady.com.

Mount Taylor, New Mexico

South / Summer / Youth

In the oratorio, when Seeker arrives at this point the *monsters* from the war begin to mature in his psyche. He remembers with clarity the bloodshed and dead bodies where he walked and witnessed the scenes of the war monster's victims.

Instead of using herbal smoke as a means to purify himself, as he would have in a ceremonial blessing, the smoke from the war pollutes him. Seeker feels survivor's guilt after his cousin brother's death and cries out against the forces that construct war. "Death, you perfect equalizer, twin brother to / war, you shiny beautiful prostitute of the powerful!" In his painful loss, Seeker understands that war is controlled and exploited by the powerful, and with this realization he is culpable, because he is part of the monster that killed his cousin brother and maimed the innocent. "The children lay like broken toys spilled on the streets / Red rags. Limbs and dreams rearranged by war." Seeker sings alone in this section without the chorus. He begins to descend further into depression and remorse. He calls on his family but is left alone without any kind of solace. Before his voice drifts off into silence, we hear: "I'm sorry! I'm sorry! I'm sorry!"

Cardinal Point Three

Figure 4.4 *San Francisco Peaks,* 2007. Source: Courtesy of Deborah O'Grady, http://www.DeborahOGrady.com.

San Francisco Peaks, Arizona

West / Fall / Adulthood

The western direction symbolizes maturity and adulthood. By now the monsters have fully matured in Seeker. He has touched death, has seen violence and bloodshed with no way to undo such destruction. Diné possess a strong aversion to making physical contact with death or placing cemeteries near the living. The only time living flowers are placed on the gravesite is at burial; thereafter, if Diné wants to leave flowers, plastic flowers will do. The belongings of the deceased are quickly disposed of after burial. References to the deceased are spoken of in the past tense—*shizhé'í Yêè* ("my father who once existed"), for example. In this way the ones who have passed are given their proper respect and people does not dwell too much on mourning, lest they might keep the spirit from making their journey into the spirit world. Seeker refutes his role as the returned hometown hero and instead wishes for a way out of his despair. And although he voices the extent of his desperation, he hasn't completely articulated what will ease his suffering.

When the story arrives at the west direction, the interplay between Seeker and the chorus takes precedence. As Seeker moves farther into the darkness, the chorus sings of the causes of his suffering as a way for him to understand his pain. They implore him with kind words, claim him as their beloved child, and try to guide him toward the light. But the war monster has entered Seeker, and he proclaims himself ironically not as his parent's child, but as a "… child of war!" More so, his language is particularly disturbing, especially when spoken in the Navajo language. Seeker sings *T'ádoo biniiyéhí'dah* ("It is a futility") in reference to his life. These words signal Seeker's profound distress. He is not in a healthy state of mind and his spirit is unbalanced; thus, there is the possibility of suicide. Later when the chorus appeals to Seeker to return to the Corn Pollen Path of life—*Hózhôjí* (the Beauty Way)—they remind him of how "Balance must always be restored" with ceremony. The chorus implores Seeker to return to the path of spiritual beauty, order, balance, and peace that are The People's ideal life purpose.

Cardinal Point Four

Figure 4.5 *Mount Hesperus*, 2007, Digital Image. Source: Courtesy of Deborah O'Grady, http://www.DeborahOGrady.com.

Mount Hesperus

North / Winter / Old Age / Death

At the end of Cardinal Point 3, Seeker finally speaks of a way out of his misery, where the *monsters* of war have fully matured. Now positioned at the northern mountain, which symbolizes winter, old age, the end of the human life cycle, and in Seeker's case, where his journey arrives at a crossroad, the chorus attempts to give Seeker the essential tools for survival, but he refuses them. They offer him the Navajo philosophy of walking on the Corn Pollen Path of life but to no avail. Seeker must make the choice to live or die. "To choose the abyss or to slay / the enemies pressed inside me." He sets out on his final mission. Now on his path toward suicide, the voices of the ancestors finally break through. The musical tempo increases and swells, creating a feeling of urgency while the chorus sings:

> Shiyázhí, remember the corn pollen path of way of life
> Your name is your shield
> Your name is your protection
> Remember your warrior name

Now at the juncture that determines his destiny, he finally accepts the ancestor's offer of a spiritual way of life in which balance and harmony are restored. The chorus arms him with virtual weapons to kill the monsters, and he systematically destroys each one. Seeker sings *fortissimo* (the Italian word and used in music for "very loud.")

> I know who I am
> I am Enemy Slayer!
> Enemy! I destroy you!
> Enemy! I destroy you!
> Enemy! I destroy you!
> Enemy! I destroy you!

As in a Navajo healing ceremony, the voices of his family and the healer sing him back into the light where healing can take place. Seeker returns from the ritual killing to reintegrate his spirit from the intensity of battle to a space where *Hózhôjí* can prevail. In doing so, he realizes his true identity is grounded as *Naabaahii* and he aligns himself with the cosmic warrior constellation and the archetypal twin, Monster Slayer, who destroyed enemies to make the world safe and was healed from PTSD. In this way Monster Slayer made himself and others safe from his destructive ways.

Having slain his enemy monsters, Seeker is purified by the ceremony and cleansed of war; he is ready to resume his return and integration into society. He can now begin to lead a more complete life where emotional affection is bonded to family and spiritually tied to the philosophy of the Diné.

> I return from the enemy
> By means of sacred prayer
> I am cleansed of war
> I am renewed with the four directions
> I am restored in a sacred manner

Return

Blanca Peak, Colorado

East / Spring / Rebirth

In the final section the oratorio returns to the east, the beginning and the rebirth of Seeker. The musical tempo slows and the chorus and Seeker sing the final lyrics together. Blessed with the power of healing words and ceremony, Seeker is transformed and restored to wellness and health. The direct experience of the ceremony is necessary for Seeker's renewal. Seeker embraces his identity and true path in life. Having been restored to *beauty*, he can return to civilian life as a different kind of *Naabaahii*, a modern-day warrior who helps others and contributes meaningfully to his community.

> May there be beauty all around
> May there be peace all around
> on earth,
> on earth.

Though *Enemy Slayer, A Navajo Oratorio* is about a returned veteran with PTSD, I do not mean to suggest that all veterans traumatized by PTSD undergo a Navajo ceremony or one similar to it. Seeker comes from a people who have always known the effects of war on the warrior's spirit and psyche and know how to treat it. However, with the alarming statistics of suicide that continue rising since onset of the wars in the Middle East, we as a country need to do more to understand PTSD and learn how to treat it so that veterans can return to civilian life and to their families as functioning members of their communities. Sadly, in 2008, as I finished writing the libretto and watched the news coverage, I was alarmed at the high suicide rate for veterans. Suicide claims more lives than warfare. Reports of death by gunfire, drownings, hangings, overdoses, and transportation accidents alarmed me. In making veterans' lives safe from the after effects of war, the United States, which called on Americans to enlist, can fulfill its responsibility to its warriors by ensuring that they are cared for and rehabilitated to prevent them turning to self-destruction and suicide. With this kind of attention and care, families, communities, and nation can be made safer. Support centers and alternative therapy have taken the first steps toward healing the veterans. Much more needs to be done. Just as importantly, we must ask what steps can be taken as a nation to prevent or minimize the loss of human life after wartime conflict.

I believe this performance gave voice to the strong oral traditions that are still living among the Diné and other Indigenous nations. With the success of this performance, I hope—and it is still my wish—that Native peoples' stories can contribute to musical performance art. Our stories—epic, secular, and entertaining—have much to contribute to all audiences and to our communities.

References

Evers, Larry. 1980. *The South Corner of Time*. Tucson: University of Arizona Press.
Maryboy, Nancy C., and David Begay. 2010. *Sharing the Skies*. Tucson, AZ: Rio Nuevo Press.
Momaday, Scott. 2001. "The Man Made of Words." In *Nothing But the Truth: An Anthology of Native American Literature*, ed. John L. Purdy and James Ruppert, 82–93. Upper Saddle River, NJ: Prentice-Hall.
Silko, Leslie. 1997. *Ceremony*. New York: Viking Press.
Tohe, Laura. 2009. *Enemy Slayer, A Navajo Oratorio*. With Mark Grey, Michael Christie, Scott Hendricks, Deborah O'Grady. © Naxos Music. Naxos 8.559604. Compact disc.

Part II

Resistance

5 Dancing at the End of the World

The Poetics of the Body in Indigenous Protest

Janet Fiskio

> . . . when I am dancing
> with my tribe during the powwow at the end of the world.
> —Sherman Alexie, "The Powwow at the End of the World"

Images of blasted mountaintops in Appalachia, fracking fields in North Dakota, and tar sands mines in Alberta flood our screens, and climate activists warn us that we are approaching a point of no return. But the operations of neoliberal capitalism that generate these wastelands have been exposed and disrupted by recent Indigenous social movements in the United States and Canada.[1] These movements employ expressive forms that combine cultural practices, direct action, and social media.[2] In this chapter I think through the meaning and significance of the body in these demonstrations, inspired by Susan Leigh Foster's insight that bodies, individual and collective, are "articulate matter" (2003, 395). I also consider what it means for bodies to be "on the street, in the square," based on the work of Judith Butler (2011). Beyond simply celebrating public demonstrations, Butler argues, we need to think about how certain bodies are policed in public space at all times (Butler 2014, 99–101). Indigenous demonstrations reveal the pervasive violence of settler colonialism, including the ways that environmental racism threatens cultural survival. Through these corporeal interventions in public and "sacred spaces" and in the sacrifice zones of global capital, Indigenous social movements articulate a vision of community in the midst of colonial violence (including the violence of climate change) based on collective action and enduring inhabitation (Kino-nda-niimi Collective 2014, 22). These demonstrations evoke the dance described in Spokane/Coeur d'Alene writer Sherman Alexie's poem, "The Powwow at the End of the World" (1996, 98).[3] This poem describes the devastating changes wrought by colonialism and holds out the promise that Indigenous peoples will survive beyond the apocalypse.

Movement

While the Healing Walk and Idle No More are demonstrations where the *movement* of bodies is particularly important, the blockades and encampments along proposed pipelines are places where the *immovability* of the body is crucial. Susan Leigh Foster describes "active stillness" in protest as a mode "of kinesthetic articulateness" (2003, 412). Foster reads the bodies at the lunch counter sit-ins of the Civil Rights movement as "Filled with kinetic potential ... not a state of non-action but rather a kind of motion" (412). Similarly, in her work on protests against the Keystone Pipeline (KXL) and other resistance movements, Kelly Klein calls this immovability a "still-activism" that challenges the functioning of neoliberal capitalism: "The still-activist's body blocks the KXL and the unencumbered movement of oil and capital it is intended to enable, breaking the flow of neoliberalism through the performance of an alternative mode of being" (Klein 2014, 211, 220; see also Maira and Sze 2012, 319). Both kinds of bodily presence—movement and stillness—perform the resistance and continuance of Indigenous sovereignty. This chapter examines the use of performance—including round dances, blockades, and walking—with a particular focus on anti-extraction and anti-pipeline actions. I use a variety of terms to describe these expressive forms, including performance, protest, action, demonstration, direct action, and ceremony. It is important to note that the term "protest" is not sufficient, because these performances are not only acts of resistance. These demonstrations draw on expressive forms, such as dancing, singing, and drumming, practices that create and support Indigenous community, sovereignty, and continuance, and thus exceed the context of settler colonialism.

The use of dance in contemporary political demonstrations is particularly resonant because of the long history of dance as "a means of asserting Native self-determination" (Shea Murphy 2007, 29). In *The People Have Never Stopped Dancing*, Jacqueline Shea Murphy describes the history of repression of Indigenous dance in the United States and Canada in the late nineteenth century. In the context of Idle No More's use of the round dance, the Ghost Dance is a particularly important event to remember. As elder John Cuthand explains, according to Cree tradition, the round dance was originally a ceremony to help heal a daughter's grief after losing her mother. Her mother returns in spirit and teaches her daughter the dance, promising that "'*when this circle is made we the ancestors will be dancing with you*'" (John Cuthand, quoted in Kino-nda-niimi Collective 2014, 24). The Ghost Dance was a movement in the 1880s that practiced the round dance as a ceremony to bring a "renewal" of the earth, which would include driving away white settlers, returning peace to Native peoples, and regenerating the ecosystem (Pratt 2005, 151). During this time Indigenous nations were engaged in a variety of forms of resistance to U.S. assimilation, especially to the 1887 Dawes Act, which functioned to dispossess Indigenous peoples of land. The U.S. government sought to repress the movement, ultimately murdering hundreds of people in the Wounded Knee massacre of 1890 (Pratt 2005, 153). By dancing in public space, Idle No More brings to light the history of Indigenous resistance and resilience in the face of U.S. and Canadian government "attempt[s] to control, through institutional discipline and punishment of Indian bodies, Indigenous peoples who resisted state authority by continuing to exist" (Shea Murphy 2007, 30).

Further, the use of the term "protest" to describe Indigenous expressive culture can serve to reinforce the current policing of Indigenous bodies. For example, after Reyna Crow, founder of Duluth Idle No More, was threatened with arrest at the Mall of America, she clarified the meaning of round dances:

> While it is true that INM [Idle No More] has organized around gravely serious causes, ... the characterization of the round dances as "protests" is not just incorrect, it's insulting. ... Not understanding is one thing. Telling a substantial segment of the community that it is unwelcome to make use of the mall—which does seem to gladly function as a sort of public square when it comes to Santa Claus and Christmas trees—to hold a brief and joyous dance with song reflecting traditional Anishinaabeg cultural values—is a message this community should be ashamed of.
>
> (Crow, quoted in "New Year's Eve," 2013)[4]

In her analysis of Tanya Lukin Linklater's (Alutiiq) *Woman and Water* performance, Shea Murphy (2010) asks, "Rather than talking about the ways that radical performance practices *challenge* or disrupt the nation state, could we think instead about what kinds of communities and political constituencies they *constitute*?" (169; emphasis in original). Shea Murphy also argues that "Indigenous dancers' bodies, despite the physical effects of colonization, are a location of ways of being and knowing, held in bodies and everyday movements. And movement practices—including contemporary movement practices—are a tool for locating and unearthing these ways of knowing" (Shea Murphy 2007, 9–10). The Indigenous actions I highlight here are constructive as well as resistant; and while they may respond to environmental racism, they are not acts of environmental activism in the popular sense. Indigenous communities do act as "defenders of the land and water," but it is crucial to note that these are acts of sovereignty, cultural survival, and decolonial struggle (Schmidt 2013).[5]

The direct action of North American Indigenous activism emerges from the 500-year struggle against settler colonialism by Indigenous peoples throughout the Western hemisphere. In her work on Simon Ortiz's (Acoma) *Fight Back*, Joni Adamson analyzes the 1680 Pueblo Revolt as an early expression of pan-tribal coalitions of resistance as well as collaboration with non-Indigenous allies (Adamson 2001, Chapter 3). In her more recent work on cosmopolitics, Adamson traces the creation of an international Indigenous movement from these early uprisings through the 2010 Cochabamba Declaration, with its advocacy for the rights of Mother Earth, tracking the emergence of a "revolutionary consciousness that accounts for hemispheric and global perspectives" (Adamson 2012, 19). In the United States and Canada, settlers have sought to dispossess Indigenous peoples through a variety of tactics, such as genocide, the establishment of reserves, and the court system. In the twentieth and twenty-first centuries, pollution from industrial projects, especially energy extraction (including uranium, coal, oil, and tar sands mining, as well as fracking for natural gas) have enacted new forms of dispossession through damage to the health of human and nonhuman beings and of ecosystems. Industrial pollution is a form of "toxic colonialism," violating treaty rights to hunt, fish, and gather on traditional lands and making areas uninhabitable or inhabitable only at great cost to Indigenous peoples' well-being (Reed 2009). Some projects, such as uranium mining and tar sands extraction, are so destructive that they can make the land unsafe for the horizon of human time.

Understanding the work of Indigenous direct action against extractive industries and pipelines is particularly important in the contemporary context of fossil fuel extraction and climate change. Through blockades, walks, and demonstrations, the bodies of protestors map the network of pipelines crossing the land and also trace the way petrochemical corporations, in conjunction with police intervention and surveillance, undermine democratic decision making. Tony Weis et al. (2014) observe that "the erosion of environmental monitoring and regulatory capacity in Canada is part of a much broader assault on democratic processes and institutions, including the muzzling of government scientists and the concealment of ostensibly public hearings" (13). Further, legislation like Bill C-51 threatens to suppress civil disobedience. Bill C-51 is presented as anti-terrorism legislation; it strengthens the powers of Canada's security state, including surveillance capacities (Cuthand 2015, Linnitt 2015). In response to this bill, Margaret Atwood (author of the classic feminist dystopian novel *The Handmaid's Tale* and the climate chaos trilogy *Maddadam*) tweeted: "#BillC51 about to be passed in Canada. See you in the slammer, kids." Bill C-51 certainly conjures images of dystopian futures, as depicted in Atwood's novels, where dissent has become a crime. Writer and documentary filmmaker Doug Cuthand (Little Pine First Nation) explains that the bill:

> applies to any activity that "undermines the sovereignty, security or territorial integrity of Canada or the lives or the security of the people of Canada." This includes the following: "Interference with the capability of the Government of Canada in relation to intelligence, defence, border operations, public safety, the administration of justice, diplomatic or consular relations, or the economic or financial stability of Canada."
>
> (Cuthand 2015, n.p.)

In addition, Carol Linnitt observes that "A Public Safety Canada Plan for Critical Infrastructure for 2014–2017 recommends increased collaboration between critical infrastructure partners and industry. The plan includes granting security clearance to oil and gas industry representatives so they can be brought in on sensitive information and secret intelligence" (Linnitt 2015, n.p.).[6] As Sunaina Maira and Julie Sze warn, "the contemporary globalization of policing as a technology of repression … is part of the transnationalization of the security state apparatus, in which models of securitization are shared and developed collaboratively among repressive nation-states" (2012, 324).

In addition to the threat of new surveillance regimes, daily life for Indigenous peoples under settler colonial governments is a situation of constant exposure to violence. The violence against the body in public is a material rendering of structural violence (Foster 2003, 396, 402; Maira and Sze 2012, 319). Protests have the power to reveal this violence and to expose the systems of White supremacy that govern public space and public life in the United States and Canada. In thinking through the condition of precarity in daily life, Judith Butler (2014) reminds us that under regimes like occupation, some bodies are exposed to violence at all times (100). Similarly, Indigenous bodies in the United States and Canada are in a state of precarity. For example, Idle No More organizes to draw attention to violence against Indigenous women, calling for investigation into the cases of murdered and missing women and girls ("Calls for Change"). Pamela Palmater (Eel River Bar First Nation) points out that First Nations people have been criminalized throughout history by the settler colonial government of Canada. She describes the ways "validly enacted laws that Canada has had that have ended up in the killing, murder, rape, violence, sterilization, and scalping of our people. Those were valid laws. The only way to protect ourselves was to act unlawfully in resistance" (Palmater 2015). This is a crucial history to remember in the context of current protest. But while Bill C-51 or other legislation may label these demonstrations as "unlawful," the *form* chosen by these activists challenges the validity of the settler nation-state. In occupations of public space, Judith Butler (2011) argues, "the body 'speaks' politically":

> the legitimacy of the state is brought into question precisely by that way of appearing in public, the body itself exercises a right that is no right; in other words, it exercises a right that is being actively contested and destroyed by military force, and which, in its resistance to force, articulates its persistence, and its right to persistence.
>
> (Butler, 2011, n.p.)

In practicing the round dance in public spaces, Idle No More reminds witnesses that dance was outlawed as part of the genocidal project of the nineteenth century, and in doing so discloses the current efforts to criminalize Indigenous ceremony as part of this same project, even as it presents a continued tradition of resistant and sovereign constitution for Indigenous peoples (Shea Murphy 2007, 30–32).

Current Indigenous activism in the United States and Canada against energy extraction and pipelines takes place within a long history of resistance (The Kino-nda-niimi Collective, 20). Glen Coulthard (Yellowknives Dene) (2014, 35) identifies the 1980s "practices of disruptive counter-sovereignty" in Canada as a particularly important context for thinking about Idle No More. He observes that this history of Indigenous direct action challenges "two pillars of colonial sovereignty": "the Canadian state's claim to hold a legitimate monopoly on use of violence and the conditions required for the ongoing accumulation of capital" (36, 37). As I am writing this essay in 2015, it is the 25th anniversary of the occupation of Kanehsatà:ke (also known as the "Oka Crisis") (Idle No More "25 Years Later"). On March 8, 1990, members of the Mohawk community constructed a blockade to protect an area called "The Pines" from development (Idle No More "25 Years Later"; Lackenbauer 2014, 168). The protest was initially nonviolent. However, after the provincial police force Sûreté de Québec was deployed on July 11, shots were fired, resulting in the death of one police officer. After this, the government sent in a military force (Belanger and Lackenbauer 2014, 7; Idle No More "25 Years Later"; Lackenbauer 2014, 175, 181). This blockade was widely documented in print media as well as television and has become a formative event of Indigenous direct action (Belanger and Lackenbauer 2014, 7–8; Lackenbauer 2014; Corrigall-Brown and Wilkes 2012). Doug Cuthand observes that that under C-51, Kahnawà:ke and many other First Nations protests could be considered acts of terrorism (Cuthand 2015).

Idle No More

It is precisely the conjunction of state and capitalist violence that is revealed by the expressive forms employed by Idle No More and tar sands, anti-frack, and pipeline protestors. Idle No More takes its name from an October 2012 teach-in planned by Sylvia McAdam (Cree), Jess Gordon (Cree/Anishinaabe), Nina Wilson (Nakota/Cree), and Sheelah Mclean in response to Bills C-38 and C-45, proposed by the Canadian government (Barker 2015, 47; Kino-nda-niimi Collective 2014, 21–22). These two pieces of legislation threatened a series of Acts that protect the environment (especially water) as well as proposing "drastic changes to the Indian Act" (Kino-nda-niimi Collective 2014, 21). Idle No More spread from this teach-in throughout North America and across the planet through "grassroots, non-hierarchical effort" (Kino-nda-niimi Collective 2014, 21–22; Barker 2015, 47).

Idle No More employs a range of expressive forms including "sharing sessions, protests, blockades, and round dances in public spaces and on the land, in our homelands, and in sacred spaces" (Kino-nda-niimi Collective 2014, 22). For example, Idle No More activists and allies took canoes out to sea to block the passage of Shell's Polar Pioneer drilling rig, which would endanger the lives and livelihoods of Alaskan Native peoples and the oceanic ecosystem (Associated Press in Seattle 2015). They were in addition protesting drilling in the Arctic, which contributes to the production of fossil fuels and thus to climate change (Associated Press in Seattle 2015). As part of the coalition of demonstrators in the #sHellNO action, Indigenous women held a ceremony, "Native Women Rising," to pray for the water and for an end to arctic drilling (Rios 2015). Indigenous resistance to tar sands mining and fracking, and to the pipelines that connect mines to global markets, has employed similar ceremonial and performative direct actions. As Foster describes in her work on the Civil Rights movement, in protests the vulnerability of the body reveals the violence of the state and of unjust laws that govern public space (2003, 396, 402). In such activism, bodies reveal that climate change is a form of what Rob Nixon calls "slow violence": "a violence that occurs gradually and out of sight, a violence of delayed destruction that is dispersed across time and space, an attritional violence that is typically not viewed as violence at all" (2011, 2). The corporeal violence against Indigenous bodies makes visible the slower, hidden violence that occurs in the poisoning of water from tar sands industry, the rise in cancers in fish and animal bodies, and the disruption of ecological cycles and the livelihoods of both human and nonhuman beings under the new climate regime.

The constellation of tactics utilized by Idle No More exceeds theories of direct action and performance studies and, indeed, is not fully legible to any academic analysis (including this chapter). These performances embody what Dean Rader (2011) calls "engaged resistance": "a fundamentally indigenous form of aesthetic discourse that engages both Native and American cultural contexts as a mode of resistance against the ubiquitous colonial tendencies of assimilation and erasure" (2011, 1). Rader describes "the ability of Native creative productions like literature and art to resist erasure, bind communities, and articulate a discourse of survivance" (2001, 1). Like literature and art, round dances and other actions function not only as critique and resistance, but also as active constructions of Indigenous space, meaning, and community that enable continuance (for a discussion of Indigenous dance, see Shea Murphy 2007). Drawing on Indigenous hip-hop aesthetics and theory, Karyn Recollet (Cree) (2015) argues that "urban flash mob round dances" function as forms of "spatial tagging" that produce "new geographies of resistance" as well as opening a "space" for "decolonial love [to] emerge" (2015, 129, 130). Recollet observes that the "impermanence and fluidity" of the round dance, as well as its significance as "a practice of relationship or alliance building," is central to the power of the dance to disrupt "colonial capital and accumulation" (2015, 132, 136). This is especially important to those actions by Idle No More that take place in public spaces and public squares, as well as the pseudo-public spaces of global capitalism and consumer exchange, such as the Mall of America (Gaard 2014). In addition, Idle No More has spread from Canadian First Nations to become a global movement of Indigenous solidarity and resistance to colonialism in all its forms.

One of Idle No More's early round dances (12 December 2012) took place in the eastern rotunda of the Mall of America, a well-recognized megamall located in Minneapolis, Minnesota (Rivas 2013). Idle No More creatively and effectively employed social media as an organizing and political tool from the outset, and therefore there is widespread documentation of this historic event. Doug Cook's YouTube video (2012) shows Indigenous peoples gathered in the mall's open space; more than one thousand Indigenous people participated (Rivas 2013).[7] The innermost circle is formed by a group of drummers. The movement of dancers around this inner circle is slow, and the chanting and singing is unfaltering, with spontaneous eruptions into cheering from the crowd. Shoppers stand transfixed on the ground floor and on a series of balconies above, their own camera phones extended to document the event.

Figure 5.1 Photograph tweeted by Ossie Michelin (@Osmich) for APTN on 9:07 A.M. 17 Oct 2013. The original tweet reads, "Tensions flare as the police line advances. Police tear gassing the crowd." Source: Courtesy of Ossie Michelin and APTN.

This round dance demonstrates what Barbara Browning calls "the political potential of the choreographic" as the dancers literally turn their backs on the storefronts of neoliberal global capitalism and engage in an enduring ceremonial practice (2004, 168). The Kino-nda-niimi Collective affirms that "In the winter of 2012–2013, our Ancestors danced with us. They were there in intersections, in shopping malls, and in front of Parliament buildings. They marched with us in protests, stood with us at blockades, and spoke through us in teach-ins" (2014, 24). The free movement of Indigenous bodies in this space testifies to the 500-year practices of resistance and continuance. The ceremony encourages and sustains the dancers, drummers, and other participants. And as both Greta Gaard and Adam Barker have observed, in the context of ongoing setter colonialism and threats to Indigenous treaty rights and life itself from toxic invasion by energy companies, the choice of the shopping malls as sites of disruption and "spatial occupations" is particularly significant (Barker 2015, 52, 55; Gaard 2014). Gaard notes that the Mall of America "is owned by the Canadian company Triple Five Group," which supports natural resource exploitation on Indigenous lands (2014). In contrast to the Best Buy banners that hang from the Mall's ceiling advertising "The Next Big Thing is Here," members of the circle carry Idle No More signs and flags of tribal nations, performing resistance in the center of colonial power as it is currently expressed through multinational corporate exploitation of Indigenous lands and the poisoning of Indigenous bodies (Rivas 2013; Cook 2012).

Elsipogtog

The round dances of Idle No More demonstrate the power of the body to choreograph anti-colonial resistance (Browning 2004, 164). Similarly, the photograph of a Mi'kmaq protestor kneeling in front of a wall of RCMP bodies while holding up a feather choreographs an immovability or "stillness" that is an active form of resistance to colonial violence, as dance scholars have articulated (Foster 2003; Klein 2014). Members of an alliance led by the Elsipogtog First Nation in New Brunswick constructed a blockade on Route 11 to prevent Texas-based SWN (Southwestern Energy) Resources Canada from conducting seismic testing (a form of exploration for shale gas reserves) on unceded land (Schmidt 2013). In Canada, unceded lands are territories "never signed away through a treaty or conquered by war" (Lukacs 2014). This blockade followed on a series of civil disobedience strategies that had blocked SWN's exploration work during the previous summer (Merrill 2014). "Protectors of the land and water" drummed and danced around burning tires while blockading (Schmidt 2013). The RCMP confronted protestors on 17 October to enforce a court injunction to allow SWN to continue their shale gas development plans (Schmidt 2013).

In Ossie Michelin's (Inuit) photograph (taken with an iPhone and tweeted [@Osmich]), Amanda Polchies kneels in a prayerful position while holding an eagle feather (Schilling 2013). She is dressed simply in a gray sweatshirt.

Polchies faces down a wall of RCMP police, who all appear to be men, and almost all appear to be White. Plastic restraints dangle from their belts; one man is holding some kind of gun (perhaps armed with rubber bullets); another seems to be wearing a bulletproof vest. The bulky dark blue jackets (perhaps covering more bulletproof vests) and overlapping positions make the men seem larger. The weapons, uniform clothing, and solid line express the power of the police in a settler-colonial regime, especially in contrast with the lone woman's body placed before them.[8] The police bodies are in active stances; they seem to be caught in midstep as they advance on the protestors we cannot see, those who are behind Polchies. She has placed her body between the RCMP and the rest of the people, stopping them in their tracks. Her nonviolent pose (on her knees, leaning forward, one hand at rest in her lap, the other holding a feather) communicates multiple meanings. Her body asks that the mass of police not hurt those behind her. Her body also says they will have to go through her to hurt those she is protecting; that she will sacrifice herself for those she loves.

The feather clearly cannot be used as a weapon, but it expresses a different kind of power than the violent potential of the police bodies. Polchies's bodily contact with the earth expresses her love for, connection with, and action to protect the earth. The feather gestures to the relations of the Mi'kmaq with nonhuman beings that long precede the violence of colonization. In this context, the feather is an expression of alliance between Indigenous peoples and their animal relations in the struggle against the forces of destruction, as well as the Mi'kmaq people's commitment to protecting the land and water on which all life depends. Through the camera's lens we see the police from the perspective of the First Nations blockaders, standing behind Polchies, not the police. Although it is a form of social media, and therefore a virtual witness, the photograph encourages us to ally ourselves with First Nations through an experience of embodied empathy: sharing the perspective of First Nations protestors. And since Polchies is closer to the camera, her body is larger than those of the police (Corrigall-Brown and Wilkes 2012, 237–38). This means that although she is a single person facing an overwhelming force, her body fills the frame, and the police are merely a backdrop. Perhaps part of the reason this photograph has spread so widely is because it articulates the power of Indigenous resistance in the face of state violence. This action by Polchies, and Indigenous protests more widely, reveal the violence of the state, the violence of energy extraction, and the violence of climate change, making the forces of structural injustice visible through the confrontation of Indigenous bodies with colonial forces (Foster 2003; Maira and Sze 2012). Further, the focus on protecting and defending air, water, and soil in Indigenous activism makes visible the violence done against other-than-human bodies; the ones some Indigenous peoples call "all our relations" (LaDuke 1999). As this photograph and the iconography it has inspired reveals, the demonstrations of Indigenous communities both testify to and enact Indigenous power and survivance as well as solidarity with nonhuman peoples in the face of multiple levels of violence.

The Healing Walk

Idle No More, anti-frack, and anti–tar sands activism confront mutually reinforcing threats to Indigenous sovereignty and cultural survival. All these struggles are linked by the extraction of natural resources, especially fossil fuels, and the exercise of national power by settler governments (whether in the United States or Canada). The Healing Walk calls attention to the devastation of the earth and First Nations communities by tar sands mining in the Athabasca River Basin.[9] The Alberta tar sands are located underneath boreal forest, which provides habitat to wildlife and subsistence hunting, fishing, and gathering of food and medicines to local Cree, Dene, and Métis communities. The scale of the mining is massive, the largest industrial project on the planet (Weis et al. 2014, 4). In addition to destroying habitat, tar sands mining pollutes groundwater, threatening human health and food security of First Nations communities (Weis et al. 2014, 9). Crystal Lameman (Beaver Lake Cree Nation) describes the multiple impacts on her community (2014, 120). These losses are spiritual and collective; as Lameman observes "My children cannot safely drink water straight from the land in the way I did as a child, and in the way my aunts, uncles, parents, and grandparents did ... They cannot fully appreciate how water has life-giving abilities" (2014, 120). Further, Lameman expresses the way the loss of subsistence practices of hunting and gathering damages collective ways of knowing and ways of life: "our oral histories, knowledge systems, and teaching are under attack ... The elderly could talk of these medicinal plants with ease, knowing where they could be found in abundance. We are losing these connections with the land" (2014, 121). Globally, the tar sands release carbon dioxide that accelerates climate destabilization, and this particularly impacts Indigenous peoples, as Clayton Thomas-Muller (Mathais Colomb Cree Nation) explains:

> The carbon footprint of the tar sands, the full life cycle of it, is driving catastrophic climate change, which of course disproportionately affects indigenous peoples and coastal-dwelling peoples all over Mother Earth, I think most directly people in the arctic, and of course people from small island states who are seeing their entire nations submerge under rising sea levels.
>
> ("TRNN Replay" 2013)

The Canadian tar sands store 240 gigatons of carbon (Hansen 2012; McKibben 2012). James Hansen of NASA has said that if all the carbon stored in the Canadian tar sands is released into the earth's atmosphere, it would mean "game over for the climate" (Hansen 2012).

The Healing Walk takes place in the center of this destruction, in an extraction site off Route 63 north of Fort McMurray in Alberta, passing a Suncor facility and then leading through Syncrude mines, processing plants, and worker housing (Thomas-Muller 2013). This is land within Treaty 8 (signed in 1899) between Canada and First Nations (Treaty 8 First Nations of Canada).

Treaty 8 is in dispute, especially in terms of the protections it offers to Indigenous ways of life including hunting, fishing, and gathering—all of which are being disrupted by tar sands mining (Treaty 8 First Nations of Canada). In addition, the Athabasca River impacts the territories of First Nations lands governed by Treaties 6, 7, and 11 (see "Healing Gathering" n.d.). The apocalyptic scale of the Alberta tar sands has been documented in a series of photojournalistic essays and documentary films that participants have freely shared through social media.[10] In addition to these visual documentations of destruction, Jesse Cardinal (2014) describes the sound of air cannons during the Walk, installed to prevent migrating waterfowl from landing in the tailing ponds after 1600 birds died this way in 2008 (131). Cardinal explains:

> Every year, more birds are returning to find their homes destroyed by more industry-access roads and industrial plants, more cleared and fractured land, more drained lakes and mining pits, and more pipelines, tailing ponds, and oil spills. It breaks my heart to imagine these living beings, flying thousands and thousands of miles to come home, only to find an industrial wasteland. (2014, 128)

The Healing Walk expresses this grief through movement. The Healing Walk was organized and led by the Keepers of the Athabasca, a coalition of "First Nations, Metis, and Inuit peoples, environmental groups, and concerned watershed citizens. and communities working together for the protection of air, water, and land, and thus, for all living things today and tomorrow in the Athabasca River and Like Watershed" ("Keepers of the Athabasca" 2016, "Our concerns and actions" page). The leadership of the local First Nations was evident throughout the event, a powerful assembly of elders and young organizers; it was also clear that women are central to the Healing Walk and to the ongoing work against tar sands destruction. The Healing Walk included a day of teaching and learning at Indian Beach on Willow Lake near Anzac, Alberta, about 50 km south of Fort McMurray ("Schedule," "Logistics," Tar Sands Healing Walk). On the 400 km drive from Edmonton, the countryside shifts from industrial farming to open fields and forest. But the evidence of the tar sands development shows in the expansion of the highway on either side of the road and the massive trucks carrying equipment and pipe. The day before the Walk was full of workshops with local elders and activists, who spoke about the impact of tar sands mining on their lives and communities. The local First Nations served a feast Saturday night.

The fifth Healing Walk in 2014 began with young women drumming and singing to wake participants.[11] The Healing Walk was led by First Nations elders, many of them women, and the walk moved at a slow and quiet pace. But this slow movement through the destruction can be exhausting; Dene drummers sustained the people. The local organizers and marshals were careful to keep walkers within the boundaries agreed to with the corporations and police. Police vehicles shadowed the walkers, but did not interfere, while some tanker truck drivers blew their horns in support.

In walking as a collective and for healing, the act of walking becomes itself a ceremony. It is essential to understand that the Healing Walk is not a typical protest or political demonstration, as the Facebook page for the event explains: "The Healing Walk was born out of a need to heal. The Healing Walk is not a rally, march, or protest, but an acknowledgement of the people and other living beings, the water, the land and the air, that is suffering due to our unhealthy energy addictions" ("Healing Gathering" 2016). Clayton Thomas-Muller (2013) describes the intentions of the Healing Walk in this way: "The idea was not to have a protest, but instead to engage in a meaningful ceremonial action to pray for the healing of Mother Earth. … This was done by turning to ceremony and asking through prayer and the physical act of walking on the earth for the hearts of those harming Mother Earth through extreme energy extraction to be healed." The act of slowly walking through this apocalyptic industrial wasteland is a compelling movement. It is a witness to the suffering of the earth. Thomas-Muller remembers:

> What I know is that a bear showed itself to us at the start of our walk and that it carried with it the teachings of courage and protection. Later, an eagle flew over us and it represented the teaching of truth and unconditional love.
>
> While we walked, we made offerings of tobacco and water on four strategic points along Highway 63. We prayed to each of the four directions and … called upon spirit, creator, mother earth, and all of the sacred elements to both heal the land and to touch the hearts, minds, and spirits of those responsible for her desecration. This was done so that the people destroying her could truly understand what they were doing. And wake up.
>
> (Thomas-Muller 2013)

Alliances

The Healing Walk is an example of coalitional politics that offers hope for further decolonial work through the power of "bodies in alliance" (Butler 2011). While the event was organized and led by local First Nations, Indigenous peoples from throughout the United States and Canada participated in a show of solidarity. In addition, the Walk has welcomed non-Indigenous allies to participate and learn. The conjunction of fossil fuel extraction and climate change is a threat to all life on earth. Indigenous movements not only protect land and water for all people; these movements, by standing on the front lines, also expose the operations of settler colonialism in the present and offer the opportunity to shift the trajectory of the future. Just as Amanda Polchies refused to be moved as she touched the earth and raised an eagle feather to the sky, Idle No More, the Healing Walk, and other Indigenous social movements stand with the powers of the land and water. Nitanis DesJarlais (Cree/Métis,

Tsimshian), reflecting on the 2014 Healing Walk, recalls a friend's observation that "we are walking with the power of the water."[12] In the 2015 elections, ten First Nations, Inuit, and Métis members of Parliament were elected to serve in Canada (McKenna 2015). Royal Dutch Shell has cancelled its plans for drilling in the Arctic (Krauss and Reed 2015). U.S. President Barack Obama has rejected the Keystone XL Pipeline (Davenport 2015). In the midst of colonial violence and the threat of climate change, this is a powerful movement to witness, and an invitation to place our "bodies in alliance" (Butler 2011).

Acknowledgments

I am grateful to the organizers and participants of the 2013 and 2014 Healing Walk. Special thanks to Ossie Michelin and APTN for generous permission to reprint his stunning photograph. Thanks also to Oberlin College for a Grant-in-Aid award that supported this project. For thoughtful comments on earlier drafts of this essay, my thanks go to Joni Adamson, Victoria Fortuna, Salma Monani, Chie Sakakibara, and Ted Toadvine. My gratitude also goes to the panelists and audience of "Cosmopolitics beyond the Anthropocene I," organized by Kyndra Turner, at the Association for the Study of Literature and Environment 2015 annual meeting. Finally, my thanks to Victoria Fortuna for the insight that Indigenous performance is a form of epistemology and her recommendation to explore the work of Jacqueline Shea Murphy, and to Chie Sakakibara for bringing Michael Rios's interview with Sweetwater Nannauck to my attention.

Notes

1. When referring to individuals, I supply tribal affiliations (when this information is available to me). I use the term "Indigenous" when referring to peoples collectively in the United States and Canada. When speaking specifically about Indigenous peoples in Canada, I use the term "First Nations."
2. I do not focus on social media in this chapter, but recommend Barker (2015). In this volume, see Chapter 6 for an analysis of an example of Indigenous web media.
3. A powwow is a gathering of Indigenous peoples that features important forms of cultural expression, including singing, drumming, dancing, and arts and crafts (Browner 2002, 1).
4. For an analysis of this event with a particular focus on gender politics, see Gaard (2014).
5. For further discussion, see the Defenders of the Land website (http://www.defendersoftheland.org/about).
6. The full RCMP report is available at DeSmogCanada: http://www.desmog.ca/2015/02/17/leaked-internal-rcmp-document-names-anti-petroleum-extremists-threat-government-industry.
7. For video footage, see Rivas 2013, http://colorlines.com/archives/2013/01/native_american_flash_mob_round_dance_takes_over_mall_of_america.html. Retrieved 1/13/2016; Cook, 2012: https://www.youtube.com/watch?v=Vn5PFHlm1ak.

8. My reading of this photograph draws inspiration from Sunaina Maira and Julie Sze's (2012) analysis of the pepper spraying of nonviolent student protestors during an Occupy protest at the University of California, Davis and from Catherine Corrigall-Brown and Rima Wilkes (2012).
9. In May 2016, after the completion of this chapter, a wildfire devastated Fort McMurray, forcing evacuations of 88,000 residents, including First Nations members ("Fort McMurray First Nation Evacuated" 2016). First Nations communities offered assistance to evacuees during the fire (Power 2016) and are now working on recovery efforts. Information on ways to support these efforts can be found at: http://www.theglobeandmail.com/news/alberta/the-fort-mcmurray-fire-heres-how-you-can-help-and-receivehelp/article29850877/; http://www.acfn.com/#!donate/qvcu8.
10. See, for example, the collection of stories at the Tar Sands Healing Walk webpage: http://www.healingwalk.org/stories.html; and this archive of photographs on Flickr, titled "Healing Walk slideshow": https://www.flickr.com/photos/healingwalk/albums/72157634550009848
11. Video footage of the 2014 Healing Walk is available at https://www.youtube.com/watch?v=OeFBW5R7xXM.
12. See https://www.youtube.com/watch?v=OeFBW5R7xXM. 3:15.

References

Adamson, Joni. 2001. *American Indian Literature, Environmental Justice, and Ecocriticism: The Middle Place*. Tucson: The University of Arizona Press.

———. 2012. "'¡Todos Somos Indios!' Revolutionary Imagination, Alternative Modernity, and Transnational Organizing in the Work of Silko, Tamez, and Anzaldúa." *The Journal of Transnational American Studies* 4 (1): 1–26.

Alexie, Sherman. 1996. "The Powwow at the End of the World." *The Summer of Black Widows*. Brooklyn, NY: Hanging Loose Press, 98.

Associated Press in Seattle. 2015. "'Shell No': Hundreds Take to Boats in Seattle to Protest Arctic Drilling." *The Guardian*. May 16. http://www.theguardian.com/business/2015/may/16/paddle-in-seattle-shell-arctic-oil-drilling.

Atwood, Margaret. 2015. (@MargaretAtwood). Twitter post. May 5, 8:20am. Accessed January 13,2016. https://twitter.com/margaretatwood/status/595609099712274432.

Barker, Adam J. 2015. "'A Direct Act of Resurgence, a Direct Act of Sovereignty': Reflections on Idle No More, Indigenous Activism, and Canadian Settler Colonialism." *Globalizations* 12 (1): 43–65.

Belanger, Yale D., and P. Whitney Lackenbauer. 2014. Introduction to *Blockades or Breakthroughs? Aboriginal Peoples Confront the Canadian State*, edited by Yale D. Belanger and P. Whitney Lackenbauer, 3–50. Montreal, Quebec: McGill-Queen's University Press.

Browner, Tara. 2002. *Heartbeat of the People: Music and Dance of the Northern Pow-wow*. Urbana: University of Illinois Press.

Browning, Barbara. 2004. "Choreographing Postcoloniality: Reflections on the Passing of Edward Said." *Dance Research Journal* 35/36 (Winter 2003–Summer 2004): 164–69.

Butler, Judith. 2011. "Bodies in Alliance and the Politics of the Street." *EIPCP: European Institute for Progressive Cultural Policies*. September. Accessed January 13, 2016. http://www.eipcp.net/transversal/1011/butler/en.

———. 2014. "Bodily Vulnerability, Coalitions, and Street Politics." *Critical Studies* 37 (1): 99–119.

"Calls for Change." *IdleNoMore*. Accessed January 7, 2016. http://www.idlenomore.ca/calls_for_change.

Cardinal, Jesse. 2014. "The Tar Sands Healing Walk." In *A Line in the Tar Sands: Struggles for Environmental Justice*, edited by Toban Black, Stephen D'Arcy, Tony Weis, and Joshua Kahn Russell, 127–33. Oakland, CA: PM Press.

Cook, Doug. 2012. "Idle No More Round Dance Mall of America." YouTube video, 5:30mins. December. Accessed January 13, 2016. https://www.youtube.com/watch?v=Vn5PFHlm1ak.

Corrigall-Brown, Catherine, and Rima Wilkes. 2012. "Picturing Protest: The Visual Framing of Collective Action by First Nations in Canada." *American Behavioral Scientist* 56 (2): 223–43.

Coulthard, Glen. 2014. "#IdleNoMore in Historical Context." In *The Winter We Danced: Voices from the Past, the Future, and the Idle No More Movement*, edited by The Kino-nda-niimi Collective, 32–36. Winnipeg, Manitoba: ARP Books.

Cuthand, Doug. 2015. "Bill C-51 Has Potential to Scoop Up Aboriginal Rights Activists." *CBC News*. May 6. Accessed January 13, 2016. http://www.cbc.ca/news/aboriginal/bill-c-51-has-potential-to-scoop-up-aboriginal-rights-activists-1.3009664.

Davenport, Coral. 2015. "Citing Climate Change, Obama Rejects Construction of Keystone XL Oil Pipeline" *New York Times*. November 6. Accessed January 12, 2016. http://www.nytimes.com/2015/11/07/us/obama-expected-to-reject-construction-of-keystone-xl-oil-pipeline.html?hp&action=click&pgtype=Homepage&module=first-column-region&region=top-news&WT.nav=top-news.

Flickr. n.d. "Healing Walk slideshow." Flickr.com. Accessed June 24, 2016. https://www.flickr.com/photos/healingwalk/albums/72157634550009848.

"Fort McMurray First Nation Evacuated as Albert Fires Grow Eightfold." 2016. Indian Country Today Media Network. May 5. Accessed June 24, 2016. https://indiancountrytodaymedianetwork.com/2016/05/05/fort-mcmurray-first-nation-evacuated-alberta-fires-grow-eightfold-164389

Foster, Susan Leigh. 2003. "Choreographies of Protest." *Theatre Journal* 55 (3): 395–412.

Gaard, Greta. 2014. "Indigenous Women, Feminism, and the Environmental Humanities." *Resilience: A Journal of the Environmental Humanities* 1 (3). Lincoln: University of Nebraska Press. doi:10.5250/resilience.1.3.007. Accessed January 13, 2016. http://www.jstor.org/stable/10.5250/resilience.1.3.007.

Hansen, James. 2012. "Game Over for the Climate." *The New York Times*. May 9. http://www.nytimes.com/2012/05/10/opinion/game-over-for-the-climate.html.

"Healing Gathering for the Land and Water." n.d. *About Healing Gathering for the Land and Water.* (Facebook.) Accessed January 7, 2016. https://www.facebook.com/tarsandshealingwalk/info/.

Idle No More Media. 2014. "Tar Sands Healing Walk 2014". YouTube video 4:13. Posted July. Accessed January 7, 2016. https://www.youtube.com/watch?v=OeFBW5R7xXM.

"Keepers of the Athabasca." *Keepers of the Water.* Accessed January 7, 2016. http://www.keepersofthewater.ca/athabasca.

Kino-nda-niimi Collective. 2014. "Idle No More: The Winter We Danced." In *The Winter We Danced: Voices from the Past, the Future, and the Idle No More Movement.* Edited by The Kino-nda-niimi Collective, 21–26. Winnipeg, Manitoba: ARP Books.

Klein, Kelly. 2014. "Disrupting the Flow: Still-Activism Redirecting Neo-Liberal Capitalism." *Choreographic Practices* 5 (2): 211–227.

Krauss, Clifford, and Stanley Reed. 2015. "Shell Exits Arctic as Slump in Oil Prices Forces Industry to Retrench." *The New York Times*. September 28. Accessed January 12, 2016. http://www.nytimes.com/2015/09/29/business/international/royal-dutch-shell-alaska-oil-exploration-halt.html.

Lackenbauer, P. Whitney. 2014. "A Bridge Too Far? The Oka Crisis." In *Blockades or Breakthroughs? Aboriginal Peoples Confront the Canadian State*, edited by Yale D. Belanger and P. Whitney Lackenbauer, 166–221. (Montreal and Kingston: McGill-Queen's University Press).

LaDuke, Winona. 1999. *All Our Relations: Native Struggles for Land and Life*. Cambridge, MA: South End Press.

Lameman, Crystal. 2014. "ᑭᐦᒋ ᐱᑭᐢᑫᐧᐃᐣ: Kihci Pikiskwewin—Speaking the Truth." In *A Line in the Tar Sands: Struggles for Environmental Justice*, edited by Toban Black, Stephen D'Arcy, Tony Weis, and Joshua Kahn Russell, 118–26. Oakland, CA: PM Press 118–126.

Linnitt, Carol. 2015. "LEAKED: Internal RCMP Document Names 'Violent Anti-Petroleum Extremists' Threat to Government and Industry." *DeSmogCanada*. February 17. Accessed January 13, 2016. http://www.desmog.ca/2015/02/17/leaked-internal-rcmp-document-names-anti-petroleum-extremists-threat-government-industry.

Lukacs, Martin. 2014. "The Indigenous Land Rights Ruling that could Transform Canada." *The Guardian* October 21. Accessed May 4, 2016. http://www.theguardian.com/environment/true-north/2014/oct/21/the-indigenous-land-rights-ruling-that-could-transform-canada.

Maira, Sunaina, and Julie Sze. 2012. "Dispatches from Pepper Spray University: Privatization, Repression, and Revolts." *American Quarterly* 64 (2): 315–30.

McKenna, Cara. 2015. "10 Indigenous Candidates Elected to Canadian Parliament, a Record." *Indian Country Today Media Network*. October 21. http://indiancountrytodaymedianetwork.com/2015/10/21/record-10-Indigenous-candidates-elected-canadian-parliament-162161.

McKibben, Bill. 2012. "Global Warming's Terrifying New Math." *Rolling Stone*. July 19. http://www.rollingstone.com/politics/news/global-warmings-terrifying-new-math-20120719.

Merrill, Stephanie. 2013. "New Brunswick Fracking Protest Broken Up with Force, Weapons and Arrests." *The Huffington Post Canada*. October 22. Accessed January 13, 2016: http://www.huffingtonpost.ca/small-change-fund/new-brunswick-fracking-pr_b_4143313.html.

Michelin, Ossie. 2013. (@Osmich). Twitter post. October 17, 9:07am. Accessed January 13, 2016: https://twitter.com/Osmich/status/390871580672135168/photo/1.

"New Year's Eve Flash Mob Round Dance Plans Spark Arrest Threat from Mall of America." 2013. *Indian Country Today Media Network*. December 27. Accessed January 13, 2016. http://indiancountrytodaymedianetwork.com/2013/12/27/new-years-eve-flash-mob-round-dance-plans-spark-arrest-threat-mall-america-152882.

Nixon, Rob. 2011. *Slow Violence and the Environmentalism of the Poor*. Cambridge, MA: Harvard University Press.

Palmater, Pamela. 2015. "'There is No Way to Save this Bill': Pamela Palmater Skewers Bill C-51." *rabble.ca*. April 9. Accessed Janaury 13, 2016. http://rabble.ca/blogs/bloggers/pamela-palmater/2015/04/there-no-way-to-save-this-bill-pamela-palmater-skewers-bill-c.

Power, Samantha. 2016. "First Nations Lend Support to Fort McMurray Fire Victims: Where to Donate." Indian Country Today Media Network. May 13. Accessed June 24, 2016. http://indiancountrytodaymedianetwork.com/2016/05/13/first-nations-lend-support-fort-mcmurray-fire-victims-where-donate-164452.

Pratt, Scott. 2005. "Wounded Knee and the Prospect of Pluralism." *Journal of Speculative Philosophy* 19 (2): 150–66.

Rader, Dean. 2011. *Engaged Resistance: American Indian Art, Literature, and Film from Alcatraz to the NMAI*. Austin: University of Texas Press.

Recollet, Karyn. 2015. "Glyphing Decolonial Love through Urban Flash Mobbing and *Walking With Our Sisters*." *Curriculum Inquiry* 45 (1): 129–45.

Reed, T. V. 2009. "Toxic Colonialism, Environmental Justice, and Native Resistance in Silko's *Almanac of the Dead*." *MELUS* 34 (2): 25–42.

Rios, Michael. 2015. "Being Idle No More: The Woman Being the Washington Movement." *Tulalip News.com*. July 8. Accessed January 13, 2016. http://www.tulalipnews.com/wp/?p=32876.

Rivas, Jorge. 2013. "Native American Flash Mob Round Dance Takes Over Mall of America." *Colorlines*. January 2. Accessed January 13, 2016. http://colorlines.com/archives/2013/01/native_american_flash_mob_round_dance_takes_over_mall_of_america.html.

Schilling, Vincent. 2013. "Woman with Eagle Feather: The Photo 'Heard' Round the World." *Indian Country Today Media Network*. November 21. http://indiancountrytodaymedianetwork.com/2013/11/21/woman-eagle-feather-photo-heard-round-world-152357.

Schmidt, Andrea. 2013. "Indigenous Canadian Fracking Protestors Refuse to Back Down." *Aljazeera America*. December 2. Accessed January 13, 2016. http://america.aljazeera.com/articles/2013/12/2/canada-anti-frackingprotestersanswerinjunctionextensionwithblock.html.

Shea Murphy, Jacqueline. 2007. *The People Have Never Stopped Dancing: Native American Modern Dance Histories*. Minneapolis: University of Minnesota Press.

———. 2010. "Gathering from Within: Indigenous Nationhood and Tanya Lukin Linklater's *Woman and Water*." *Theatre Research International* 35 (2): 165–71.

"Tar Sands Healing Walk." Tar Sands Healing Walk. Accessed June 26, 2016. http://www.healingwalk.org/stories.html

Thomas-Muller, Clayton. 2013. "A Healing Walk through Canada's Tar Sands Dystopia." *Yes! Magazine*. June 6. http://www.yesmagazine.org/planet/a-healing-walk-through-canada-s-tar-sands-dystopia?ica=Tweet&icl=ShareBar_Art_UR.

Treaty 8 First Nations of Alberta. (1899) "Articles of Treaty 8." Accessed Janaury 13, 2016. http://www.treaty8.ca.

"TRNN REPLAY: Alberta Regulator Approves Tar Sands Expansion Despite 'Irreversible' Eco Impact." 2013. *The Real News Network*. August 26. Accessed Janaury 13, 2016. http://therealnews.com/t2/index.php?option=com_content&task=view&id=31&Itemid=74&jumival=10493.

"25 Years Later, Impacts of the Oka Crisis—The 1990 Occupation of Kanehsatà:ke and Kahnawà:ke." *IdleNoMore*. Accessed January 7, 2016. http://www.idlenomore.ca/25.

Weis, Tony, Toban Black, Stephen D'Arcy, and Joshua Kahn Russell. 2014. "Introduction: Drawing a Line in the Tar Sands." In *A Line in the Tar Sands: Struggles for Environmental Justice*. ed. Toban Black, Stephen D'Arcy, Tony Weis, and Joshua Kahn Russell. Oakland, CA: PM Press. 1–22.

6 New Media, Activism, and Indigenous Environmental Governance

Politics and the Minnesota-Wisconsin Wolf Hunt

Clint Carroll and Angelica Lawson

Over the past two decades, numerous studies have assessed the role the Internet has played in activist movements (Alia 2009, 2011; Ginsburg 2008; Kahn and Kellner 2004; Loft and Swanson 2014). Likewise, scholars have focused on how Indigenous communities have "appropriated media technologies to serve their own cultural, political, and social visions" (Srinivasan 2006, 497). This chapter looks specifically at how Ojibwe activists and their communities have used the Internet, specifically the website maiingan.org, to articulate their opposition to state-sanctioned wolf hunts in the north-central U.S. states of Minnesota and Wisconsin. To contextualize the rhetoric of the former website, we accompany this analysis of new media and activism with an analysis of Indigenous environmental governance—specifically, how American Indian nations struggle to assert their traditional perspectives toward nonhuman beings *as relatives* against a dominant resource management paradigm that sees nonhuman beings in instrumental and hierarchical ways (LaDuke 1999).

In January, 2012, federal U.S. officials delisted the gray wolf (*Canis lupus*) in the western Great Lakes region from its protection under the Endangered Species Act, delegating wolf management to tribal and state authorities. Soon after, without consultation with tribes, Wisconsin and Minnesota lawmakers announced plans to authorize a wolf hunt during the fall of 2012 (October 15 and November 3, respectively). The authorization of the Minnesota wolf hunt required the legislative removal in summer 2011 of a five-year moratorium on wolf hunting contained in the state's original Wolf Management Plan (Minnesota Department of Natural Resources 2001, 2). The moratorium, which was designed to provide time to monitor wolf populations after their imminent delisting, had been the result of an extensive public input process that included state resource managers, state legislators, tribal representatives, and the general public (Edberg 2012). In a series of statements from regional tribal governments and individuals, local American Indians officially opposed the hunt, with one Bad River Ojibwe citizen saying, "Killing a wolf is like killing a brother" (Pember 2012a).

Additionally, numerous Ojibwe bands in Minnesota and Wisconsin declared their reservations as sanctuaries in an effort to protect wolves from hunting and trapping.

Connected with this movement and the activism associated with it was the website maiingan.org (*ma'iingan* is the Ojibwe word for "wolf"). This website contained information about the Ojibwe protests as well as news articles and declarations made by various Ojibwe bands. The website, now closed, serves as an example of cultural and political sovereignty through its use of Indigenous language, imagery, tribal stories, and through declarations of self-determination and documentation of sovereign actions detailed in the various links throughout the site. In addition, maiingan.org became more than just a place to connect activists and educate the public about the controversy over the wolf hunt issue. It also served as a place to articulate a framework for "relationship-based" approaches to environmental governance that are found within many Indigenous traditional values and knowledge systems (Carroll 2015, 7–9). This framework emerges from the Ojibwe people's story, which outlines their relationship to wolves as sibling beings and develops into a "cosmopolitics" via the connections made to treaty rights and the federal trust relationship with the U.S. government.

We begin our discussion by framing the wolf hunt issue within Indigenous politics, which complicate simplistic and stereotypical understandings of the "ecological Indian," and articulate instead the more complex idea of cosmopolitics. We turn next to a close analysis of the website maiingan.org to demonstrate how it placed Ojibwe cosmological views alongside political responses to the language of settler governance. As we then argue, this mesh of traditional cosmologies and contemporary legal understandings helped to demonstrate the principles of a tribal environmental governance that takes into account cultural, spiritual, and environmental health to push back against dominant, state-sanctioned ideas of resource management, which often tend to devalue both Indigenous and wolf populations.

Representation and Indigenous Environmental Politics

The field of Indigenous environmental studies is plagued with persistent stereotypes that can both undermine and bolster Indigenous claims and assertions. For example, Indigenous perspectives have been discredited by skeptics who claim that many notions of Indigenous environmentalism are romanticized myths (e.g., Krech 1999). An opposing perspective is that Indigenous people employ the discourse of "ecological nobility" in order to secure seats at the environmental decision-making table—an expression of "ecological legitimacy" (Pulido 1998, 121; Ranco 2007, 42). Based on his work with people of the Kluane First Nation, anthropologist Paul Nadasdy (2005) frames the debate over Indigenous ecological nobility in terms of a presumed "spectrum of environmentalism" on which Indigenous perspectives may or may not intersect. To Nadasdy, the construction of the spectrum as a color-coded, taken-for-granted, and two-dimensional field on which humans occupy various positions—from the "brown" end of the spectrum (anti-environmentalists) to the "dark green" end (radical environmentalists)—presumes that American Indian traditional philosophies can be placed at a single position (often at the dark-green end of the spectrum). And yet, American Indians often display complex, multidimensional points of view on "the environment" that do not fit neatly within this construction. Nadasdy argues that American Indian environmental perspectives "are more properly understood in relation to their own very different set of cultural ideals, which, when interpreted from a EuroAmerican perspective, can seem to stand in stark contradiction to one another" (321). In this light, Indigenous actions or perspectives can presumably fall on multiple points of the EuroWestern "spectrum of environmentalism" and still be consistent with traditional tribal value systems.

Adding to this discussion, anthropologist Anna Willow contends that the discourse of ecological nobility often masks the politics of what is at stake in Indigenous environmental activism. Through her work on the Grassy Narrows First Nation's blockade against the clear-cutting of forests on their traditional territory, she asserts that "when Indigenous activism [possesses] significant environmental components, it often also makes critical political statements about the inequalities that divide Indigenous communities from the nation-states and non-Native societies that surround them" (Willow 2009, 36). In common with Nadasdy's and Willow's work, the Ojibwe opposition to state-sanctioned wolf hunts points to the environmental politics of settler colonialism, in which Indigenous environmental issues are inseparable from historical and present-day political struggles. As we discuss below, the interrelated struggles over representation, wildlife policy, jurisdictional authority, and community well-being underscore the notion that, for Indigenous nations that have been subsumed by settler societies, "the *political* is *environmental*" (Carroll 2015, 12). This assertion that Native political conditions are fundamentally tied to land-based struggles rests upon many other works that have illuminated the nuances of environmentally driven Native activism and policy battles (e.g., Adamson 2001; Clark 2002; LaDuke 1999; Martin 2012; Robyn 2002; Tsosie 2007; Whyte 2013; Willow 2009; Wolfley 1999).

Despite the desire of many Native activists to earnestly convey tribal perspectives rooted in traditional philosophies, simplistic notions of "ecological nobility" often prevail in media coverage of Native environmental issues (see, e.g., Smithers 2015). The same could be said for uninformed interpretations of Ojibwe activism against the wolf hunt. Taken out of context, the Bad River Ojibwe citizen's remark, "Killing a wolf is like killing a brother" (cited above), coupled with the continued use of variations of this refrain on maiingan.org, may have evoked "ecological Indian" imagery, potentially labeling the Ojibwe perspective as an opportunistic use of this mystical stereotype. Yet, as Penobscot scholar Darren Ranco notes, to interpret

Indigenous perspectives toward the environment as simply strategic is to marginalize, silence, and ultimately disregard the cultural worldviews that ground Indigenous ecological relations (2006). In contrast, to understand the phrase "the wolf is my brother" as localized and anchored within the context of Ojibwe cosmology, we see a more complex pushback against the history and ongoing practice of settler colonialism that continues to discredit tribal regulatory authority over traditional lands and to inhibit the ability of Indigenous peoples to uphold relationship responsibilities with other-than-human beings (in this case, wolves).[1]

The concept of upholding relationships with other-than-humans is one of the centerpieces of Ojibwe philosophy (Hallowell 1975), and Ojibwe elders refer to this as natural law (Borrows 1997). Such a cosmology articulates ethical ways of being in the world. The website maiingan.org expanded upon this cosmology via images, stories, official documents, and declarations.

On the home page of maiingan.org, an image of a wolf gazed directly at the viewer next to text in Ojibwe, which was also given in English: "Where Indians' and Wolves' Paths Cross." Immediately below and to the left the text was the quote, "The wolf is my brother." Underneath this was written, "The cultural, spiritual, and historical relationship between Ojibwe Anishnaabe and Ma'iingan of the Great Lakes Region of North America." The seemingly simple statement, "The wolf is my brother," has in fact deep roots in the Ojibwe story of creation, which in turn references an entire worldview mapped out through the story cycle it initiates: "Anishinaabeg stories are embedded in relationships and relationship-making practices—they institute them [the practices], explain them, and/or define them. Many see stories as the living strands (indeed, even living beings themselves) that constitute the relationships Anishinaabeg hold between themselves and with all of creation" (Doerfler, Sinclair, and Stark 2013, 59). Thus, the content and imagery of the site expressed an Ojibwe cosmology that underscored very real connections and concerns regarding the "the cultural, spiritual, and historic relationship" between wolves and the Ojibwe people.

When such relationships between Indigenous peoples and other-than-human beings challenge Western concepts of "natural resources" politically, they evoke an "Indigenous cosmopolitics." In her work on Indigenous cosmopolitics in the Andes, Marisol de la Cadena analysed the "political reconfiguration that is currently taking place in Latin America," while focusing on "Indigenous social movements" as a critical aspect of these transformations (2012, 334). Of particular note is her analysis of the rights of nature in official Indigenous documents, arguing that the presence of "earth-beings" on the political stage in the Andes disrupts "the monopoly of [Western] science to define 'Nature'" (346). Drawing from the work of Bruno Latour (1993), de la Cadena highlights how the European Age of Enlightenment produced a particular "regime of life that created a single natural order and separated it from the social by creating an ontological distinction between things and humans that it purported universal" (342).

The "presence of earth-beings in social protests" (343) thus articulates an Indigenous cosmopolitics that disrupts this deeply rooted theory of politics, which separates man from nature and has been wielded as a universal truth.

Building on de la Cadena's work, Joni Adamson argues that "Indigenous peoples had been articulating cosmovisions of worlds more multiple than those of 'modern' nation-states for many decades, even for hundreds of years" (2014, 185). Adamson notes, "'ancient ancestral knowledges' can be useful in illuminating 'things' in the natural world that cannot be seen by the human eye, including multi-scale relationships between species functioning in systems that heretofore have not been considered deserving of the same legal rights and protections as humans" (2014, 181–82). Such Indigenous perspectives assert that human beings are a vital component of, rather than artificially separated from, the rest of creation. To acknowledge the vitality of these co-existent relations is to push back against the histories of (neo)colonial-enforced marginalization of both Indigenous people and their other-than-human relatives.

Thus, relaying the message of "wolves as brothers" through a political website encouraged site visitors to engage with Ojibwe bands on issues of natural resource management policy and served as a cosmopolitical tool, one that specifically articulated a distinct approach of Indigenous environmental governance. In demonstrating how, we might also consider the particularly powerful decolonial role new media technologies such as websites can play in Indigenous activism.

Self-representation in the media by Indigenous people the world over has been an exciting movement empowering people who have historically been rendered powerless or invisible (Ginsburg 2008; Alia 2009; Loft and Swanson 2014). With the surge in new media creation by Indigenous people, we see more accurate historical and contemporary representation as well as counter-narratives that intervene in colonial oppressive histories. Such media may serve as activist documents to be shared with audiences worldwide. In writing about the positive aspects of Indigenous new media, Faye Ginsburg notes, "... [cultural activists] see media as a means of furthering social and political transformation by inserting their own stories into international narratives as part of ongoing struggles for aboriginal recognition and self-determination (2008, 302). The Ojibwe assertion of political and cultural sovereignty in conjunction with the cosmopolitical argument articulated on maiingan.org was a form of new media activism.

Websites in particular have become a critical space for both the distribution and promotion of Indigenous media. By analyzing the website content of maiingan.org through the lens of an Indigenous "media cosmology," which "embraces an Indigenous view of media and its attendant processes that incorporates language, culture, technology, land, spirituality, and histories" (Loft 2014, xvi), we can see how Ojibwe people articulated a cosmopolitical relationship between themselves and wolves. As such, maiingan.org became a "scene [or site] of Indigenous instruction" (Allen 2002, 184). The website functioned as a "cosmological intellectual ecosystem" (Loft 2014, xvi), and served as an important site of Indigenous agency and pedagogy.

Ojibwe Cosmology, Visual Cosmopolitics, and Internet Activism

The Ojibwe creation story articulates a cosmological relationship critical to the argument for the protection of wolves. In this story, Ma'iingan was sent by the Creator to walk the Earth as a brother to Original Man. The Creator instructed them to journey throughout the Ojibwe homelands and to establish relationships with all the plants, animals, and places by visiting and naming them. In the course of this journey, Wolf and Original Man became very close to one another and developed great mutual respect. Upon completion of their task, the Creator instructed them to part ways and assume separate paths, but with the knowledge that whatever happens to one will also happen to the other. In this way, wolves and Ojibwe people were to live separately yet were forever bound to one another through a relationship of brotherhood (Benton-Banai 1988). This story relays an important cosmological relationship as expressed through Ojibwe teachings handed down from generation to generation.

In discussing Ojibwe oral traditions and the significance of supernatural beings (such as Nanabozho/Original Man) and animals, Jean Buffalo argues that the stories suggest:

> alternative ways of being in the world. The Ojibwe have rejected the notion that humans have fallen from the cosmos into existence in this world. The Anishnabeg worldview presents not that which we study of nature, but one in which we open the category of personhood to include nonhuman beings. The complexity with which the Ojibwe have articulated this way of being in the world assists us in understanding what can be described as a moral sense of nature. By their mapping of the cosmos, Ojibwe knew from whence they had come, and they know where they are going. ... (2007, 40)

A printable poster displayed on the website maiingan.org presents a visual representation of this cosmological relationship of wolf as brother (Figure 6.1). The poster features Nanabozho and three wolves with the heading "*Niikaanag*: My Brothers, My Friends," and includes a brief version of the Wolf and Original Man story printed below the image. The artwork for the poster was by renowned Ojibwe artist Rabbett Strickland, and it became a symbol for the movement; it was available on the website and distributed at conferences, such as the White Earth Wolf Conference in 2013. In Strickland's painting, Nanabozho, or Original Man, is sitting with three wolves in a muted landscape. Nanabozho's image serves as a symbolic reminder for those viewers literate in Ojibwe creation stories, because it was Nanabozho who taught the people how to live ethically in a multispecies cosmos: "He behaved more like a human being than a Manitou. He was sent to the world by Kitchi Manitou to teach the Anishnabeg ... He became the messenger ... an emissary on earth between the different species of beings, and a protector for the Anishnabeg"

(Buffalo 2007, 27). As stated earlier, Original Man (Nanabozho) was given instructions to walk the earth with Wolf and form a relationship with all living things. Thus the story, for which the painting provides a visual reminder, articulates an Ojibwe cosmology in which wolf and man are part of a multispecies cosmos for which they have responsibility.

Figure 6.1 *Niikaanag:* My Brothers, My Friends by Rabbett Strickland.
Source: Courtesy of Rabbett Strickland.

The composite imagery for this poster is significantly different from many of Strickland's other works. Typically, a Strickland painting is massive in scale, uses vibrant colors, and contains "hidden meanings and messages, icons both familiar and strange … he eschews the typical sentimental reverence and immediate access of typical trickster insouciance, and instead embeds a host of references and symbols deep into the life of his paintings" (Treuer 2008, 20). The wolf brother painting does not contain hidden meanings but is rather quite literal in the tonal composition and blending of wolf, man, and landscape. The painting's imagery is straightforward, but gathers accumulative meaning in its literal layering of earth tones and figures. Nanabozho sits at a river's edge on a bank the same color as his skin and leans against birch bark trees that are a lighter shade than his complexion. The wolves surrounding him blend with the landscape rendered in colors that resemble rocks and earth in the same tones as Nanabozho and the trees. Two of the wolves are layered over the image of Nanabozo and a third is connected via Nanabozho's legs, which create a bridge from the bank to the river where the wolf stands with his feet submerged in water. The image, though not realistic in terms of dimension and scale, is rather stunning and surreal. More importantly, the visual message is clear. The figures are literally connected, layered over one another in similar colors, visually representing the imagery's blurb that we are all connected. Through its straightforwardness, the painting represents and provides a visualization of an Ojibwe cosmology.

The simplicity of the painting, in contrast to Strickland's other large-scale allegorical works, is indicative of the pedagogical goals of the poster. Whereas some of Strickland's more complex paintings, such as "Creation of the Morning Star" or "Moving the Sun," require that the painting be "read, not simply seen … then re-read and re-read in order to slowly, gradually, pry meaning from wonder" (Treuer 2008, 20), the wolf-brother painting is less complicated. As such, it successfully serves as a sign or symbol for the movement. The combination of the painting and the stories contained within the website helped to provide a deeper, multilayered reading of the image in conjunction with the wolf issue. The painting served as visual reminder of the cosmological relationship between Original Man and Wolf, while the embedded stories, legal documents, and declarations on the website added further depth of cosmopolitical complexity. Specifically, beyond rhetoric or a good story about wolves as brothers, the site demonstrated concrete actions taken by Ojibwe bands grounded in the ethics and responsibilities articulated in their cosmology. The website provided examples of how Ojibwe bands could assert their sovereignty and listed specific examples of actions taken. For example, embedded in the link "call or write a letter to your Minnesota state legislator now" were the following statements:

> The tribes of Minnesota refuse to allow the state to impose further colonization of Native people by non-Native people who come to our reservations for sport-killing of a critical cornerstone of our spiritual structure. We refuse to allow this act of racism and ignorance to be imposed on us by the state of Minnesota.

> The intentional act of killing wolves on Indian reservations and intentional disregard for our culture is a blatant, government-sanctioned act against our religion and culture. The state of Minnesota is in clear violation of tribal sovereignty, self-determination, human and civil rights of Native people. (maiingan.org, n.p.)

Associated with these statements were links to documents filed by various Ojibwe bands that detailed political and sovereign actions they had taken, including the "*Ma'iingan* Protection Ordinance" (Fond du Lac Business Committee 2012) and the Declaration Designating the White Earth Reservation as a *Ma'iingan* (Wolf) Sanctuary ("Declaration Designating," 2012). These documents recorded and provided evidence of "activist events" staged by the various bands (Adamson 2012; Allen 2002) and also articulated the means by which the Ojibwe could assert their treaty-based sovereign rights. Indeed, the website was a virtual online "staging" of Indigenous activism connected to physical activist events, which, as we describe below, points to the complex realities of negotiating life in settler-colonial systems.

Indigenous Environmental Governance and the Language of "Resources"

The images and stories that articulated the concept of wolf-as-relative throughout the *maiingan.org* website connect Ojibwe natural law with the official stances of numerous federally-recognized Ojibwe bands who have declared their reservations as wolf sanctuaries. Tribal ordinances and declarations that served to protect wolves against U.S. state-sanctioned hunting seasons were directly related to local concerns about the impact the hunts have on the health and well-being of Ojibwe people. Here we broadly construe "health and well-being" to include not just clinical health concerns but also cultural and spiritual aspects of health. An institutionally sanctioned wolf hunt—in its symbolic political stance and as a literal threat to wolves' lives—sends to Ojibwe people the message that their creation stories and cultural teachings are irrelevant in the context of state "resource management." Such a project, as the Minnesota Department of Natural Resources (DNR) commissioner asserted, must be an objective undertaking, informed solely by Western science and precluding consideration of "cultural arguments" (Kraker 2012).

Implications for the cultural health of Ojibwe people are profound. For example, how might Ojibwe youth reconcile, on the one hand, the stories of their elders that speak of *ma'iingan* as a brother whose fate is linked to that of the people, with on the other hand, the sanctioned hunting of wolves for sport? And what of culturally based subsistence practices that rely on healthy ecosystems, of which the gray wolf is a keystone species? The reduction of a top predator species has been shown to lead to ensuing changes that compromise overall ecosystem health, the process of which ecologists call a "trophic cascade" (Eisenberg 2010). Given the scenario of a trophic cascade caused by a decline in the wolf population due to hunting, possible resulting negative effects on plant and animal species connected to Ojibwe culture and subsistence ways threaten what Potawatomie scholar Kyle Powys Whyte terms tribal "collective continuance," or "a community's capacity to be adaptive in ways sufficient for the livelihoods of its members to flourish into the future" (2013, 518). Drawing such connections between politics, the environment, and health is essential for combating claims against "cultural arguments" in wolf management. Additionally, illuminating resource management as a product of a dominant Western paradigm exposes the inherent, but unacknowledged, "culture" of such agencies (Nadasdy 2011; see also Carroll 2013).

Given these restrictions on the "cultural" arguments and language the Obijwe can deploy, like many American Indian nations that seek to uphold relationships to the nonhuman world that are central to their tribal identities and self-determination strategies, they work with natural resource management language in order to protect and reassert control over tribal lands. Carroll has expanded on this blended strategy elsewhere in his work with the Cherokee Nation, asserting that in Native North America, forms of environmental governance "[operate] within a spectrum of *resource-based* and *relationship-based* approaches" (2015, 7–8). This strategy aims to show how, in reconciling resource control with traditional teachings that seek to uphold unique tribal relationships with the land and all life, American Indian nations defy an assumed trade-off between the trappings of a Western resource management paradigm and the tenets of their unique tribal cultures.[2] It follows that Indigenous environmental governance—in its ability to articulate both dominant discourses of "resources" and traditional understandings of land-based relationships—adds complexity to the landscape of settler-colonial environmental politics.

Furthermore, as Whyte illustrates, tribes seek to maintain not just relationships, but "*relational responsibilities*" toward nonhuman nature that acknowledge "systems of responsibilities" not readily apparent from a Western perspective (2013, 519, emphasis in original). Thus, rather than a focus on a single species, relationship-based approaches to environmental governance acknowledge the need for upholding community in its broadest sense. For example, if Wolf is missing or vulnerable, the whole community of beings is lessened. In this light, collective continuance and the "future flourishing of tribal livelihoods" encompass the maintenance of relationships with all creation and the subsequent imperative of making good on the responsibilities that come with such relationships.

The responses of tribal natural resource managers in Wisconsin and Minnesota illustrated these complex points. Specifically, tribal DNRs utilized the language of "resources" to uphold their relationships with the wolf, and this is best seen in the issue of wolf quotas and hunting license distribution. In Wisconsin, Ojibwe bands stressed their treaty rights to off-reservation resources, claiming that they have a share in the number of wolves to be legally killed (the state limit was set at 201 wolves), and that they can manage this share in any way they want, including by refusing to kill the wolves allotted to them (Erickson and Stark 2012/2013; Kraker 2012; Pember 2012b).

This "resource rhetoric" position taken by Wisconsin Ojibwe bands reveals their attempts to comply with the cultural bias inherent to state resource management policy, which favors state claims to scientific and professional objectivity (refer back to the Minnesota DNR stance of not considering "cultural arguments").[3] At the same time, because the legal settler-colonial framing of "wolf management" is somewhat distinct from other "resource management" frames (for example, the cut-and-dried resource use of game animals like deer), the Ojibwe can claim a portion of state wolf quotas via retained treaty rights *and choose to not kill/use the resource.*[4] In effect, they can take advantage of the logical inconsistencies of the state's resource paradigm to protect the wolf, a being they recognize as a literal relative. This meld of environmental governance speaks to both traditional cosmological understandings and the language of settler-colonial systems with which Obijwe bands must engage.

Other Ojibwe voices tempered the debate over the act of killing a wolf. As a way of acknowledging the complexity of human-wolf interactions in the present day, White Earth Ojibwe citizen Robert Shimek stated that controlling individual "problem wolves" was not necessarily at odds with Ojibwe values and practices ("Concerns over Wolf Hunt" 2012). While Shimek maintained adamant opposition to the state-sanctioned hunt, his reference to problem wolves sought to address the concerns of wolf hunt proponents—largely non-Indigenous farmers, ranchers, and landowners throughout rural Minnesota and Wisconsin who see wolves as threats to livestock and livelihood. Here it is worth stating that while majority views tend to dominate the discussion and media coverage, there are significant divergences within the settler/Indigenous binary that complicate the narrative. For one, not all tribes are, or have been, pro-wolf, as seen in Steve Pavlik's work on Apache tribal oppositions to the reintroduction of the Mexican wolf in Arizona and New Mexico (Pavlik 1999). And while wolves typically trigger opposition from rural, White ranchers and farmers (Fritts et al. 2003), this group is not always anti-wolf, as Tom Lynch's 2005 volume has shown with regard to the Mexican wolf reintroduction program in the U.S. southwest.

Although the nuance within Indigenous and settler perspectives on wolves challenges simplistic and dichotomous representations, at stake in the Ojibwe struggle to represent and protect wolves in the western Great Lakes was, as Whyte asserts, a "forward-looking framework" of justice that require[d] state resource managers to "*shelter* and even *amend* the persisting and emerging systems of responsibilities that constitute collective continuance" (2013, 525, emphasis in original). State-sponsored wolf hunts are only one of many ways to approach the task known as wolf management. By dismissing Ojibwe perspectives on wolves on the basis of not considering "cultural arguments," Minnesota and Wisconsin added to the "coupled political obstructions and ecological challenges" for Ojibwe collective continuance (Whyte 2013, 521). In common with Whyte, we argue that accounting for Indigenous environmental perspectives in settler-colonial contexts and grounding those perspectives in the political relationship that Indigenous nations have with settler governments opens up new avenues for understanding environmental politics and human-nature relationships. Doing so also fosters rich analyses of tribal natural resource management approaches, as well as strategies for promoting increased tribal regulatory authority over off-reservation resources. These concepts were similarly articulated on the maiingan.org website, to which we now return.

Internet Activism as Decolonial "Media Cosmology"

Although no longer an active website, maiingan.org served as a place to articulate a cosmopolitical framework for Indigenous media activism. The framework was grounded in Ojibwe worldview and was extended to legal and political mechanisms for resisting the wolf hunt. As an instrument of decolonial "media cosmology," which Steven Loft describes as an "Indigenous view of media and its attendant processes that incorporates language, culture, technology, land, spirituality, and histories" (Loft 2014, xvi), the website distilled and articulated the tenets of Ojibwe environmental governance to advocate for the rights of nonhuman beings. More than just advocacy for nonhumans, Ojibwe Internet activism against the wolf hunt as expressed by the wolf-as-brother story sought to maintain a sacred sense of balance by upholding critical "relational responsibilities" (Whyte 2013, 519) that, in turn, reinforced Ojibwe people's own presence and role in this cosmological community. As referenced in several places on the website, the Ojibwe creation history describes how their fates are intertwined, and the similarities between historical wolf policy and Ojibwe experiences with European settlers have not gone unnoticed. In *The Mishomis Book*, traditionalist Edward Benton-Banai, an Ojibwe citizen from Lac Courte Oreilles, describes this observation:

> What the Grandfather [Creator] said to them has come true. Both the Indian and the wolf have come to be alike and have experienced the same thing. Both of them mate for life. Both have a clan system and a tribe. Both have had their land taken from them. Both have been hunted for their hair. And both have been pushed very close to destruction. We can tell about our future as Indian people by looking at the wolf.
>
> (Benton-Banai 1988, 8)

Benton-Banai's tribal-specific perspective, grounded in sacred history and recent historical experience eludes any simplistic accusations of "ecological Indianness" and implores non-Indigenous citizens and state resource managers to recognize the historical and moral implications of a sanctioned wolf hunt to Ojibwe people. While the unilateral *imposition* of Indigenous values upon settler societies is not a practical nor a desirable scenario, many have argued for framing environmental issues within the federal trust responsibility toward American Indian nations, and, by extension, implementing Indigenous values into co-management agreements (see Ross et al. 2011; Whyte 2013). Taking seriously Indigenous knowledge and values as articulated through Indigenous cosmologies is both a *moral* and a *legal* issue.

The idea that the dominant settler society—as represented by the Minnesota and Wisconsin DNRs—should reevaluate its moral stance against wolves and in the process develop a new relationship with them is, in fact, not far from what prominent wolf biologists also propose. Minnesota-based wolf specialist L. David Mech argues for learning to live with wolves through modifying farming and ranching practices (this could involve reinstating traditional European practices that inhibited livestock depredation by wolves, such as guard dogs and shepherds), increasing public education (countering narratives of "the big bad wolf"), and focusing wolf control efforts on the previously mentioned "problem wolves" (Mech and Boitani 2003). For American Indians, however, moral grounds alone do not adequately address the depth of the issue, and Ojibwe bands explored many potential legal and political avenues in order to strategically inhibit the extent of the hunt. As stated above, Ojibwe bands have declared their reservation boundaries as sanctuaries. Although in many cases reservation lands are severely checkerboarded with non-Indian private lands (therefore leaving tribes with fractionated regulatory authority), tribal citizens have picketed reservation borders to persuade potential hunters to respect tribal policy against the wolf hunt (Kraker 2012).

Maiingan.org is no longer active, but its presence is still visible through a network of other media sites.[5] Its absence speaks to the ephemerality of the Internet, which has implications for activist movements that employ this tool. Maintaining such a presence cannot escape the reality of website fees, maintenance time, and technical know-how. Nevertheless, recent victories may explain the site's withdrawal, and thus how its founders may have viewed its purpose served. In December 2014, U.S. District Judge Beryl Howell overturned the federal decision to remove the gray wolf population from the endangered species list in the western Great Lakes region. Stating that the 2012 removal was "arbitrary and capricious," Judge Howell's ruling restored the ban on wolf hunting and trapping in Minnesota, Wisconsin, and Michigan (Flesher 2014). At the time of writing, wolves in the Western Great Lakes remain on the endangered species list. Both Ojibwe people and wolves can take comfort in this for now, but undoubtedly challenges will continue to arise with regard to wolf management in the western Great Lakes.

In sum, by asserting a relationship-based approach to environmental governance and through the strategic use of media technology to provide a platform for tribal resistance, Ojibwe people acknowledged the rights of other-than-human beings as expressed through traditional teachings, or cosmovisions. As seen in other Ojibwe activist sites like protectourmanoomin.org, cosmopolitical work via new media continues to promote Indigenous ways of life that value "respect over control, and physical and spiritual sustenance over profit margins" (Carroll 2015, 173). We have hoped to highlight nuances in such work that defy simplistic portrayals of Native environmental management and activism. For tribal natural resource management and Internet activism alike, the ability to account for both the necessity of resource-based approaches within a settler-colonial setting and the imperative of relationship-based approaches as articulated in Indigenous cosmovisions is a critical strategy for achieving cultural, spiritual, and environmental health. Navigating such terrain in social, political, institutional, and technological contexts presents both steep challenges and promising opportunities for Indigenous nations today.

Notes

1. Here we invoke Patrick Wolfe's work on settler colonialism, defined as "a structure, not an event" (1999, 2).
2. Here we reference the "Weberian dilemma" that Ronald Niezen (2003, 140–41) puts forth as a term to describe the presumed struggle that Indigenous peoples face between maintaining cultural traditions and developing Western-based political structures. While Niezen tends to view this struggle as a zero-sum game, we have observed tribal solutions to be much more complex.
3. Terms that state resource managers use to describe the wolf hunt include "harvest," "cull," "control," "manage," and so on; these are all terms used to describe "resources" according to Western management paradigms and the binary construction of nature/society (Nadasdy 2011).
4. Aside from selling/collecting wolf pelts, humans do not eat wolves, like they do other game animals. Thus, resource use is quite distinct.
5. Maiingan.org is no longer active, but its presence is still visible through a network of other media sites. See, for example, the White Earth Land Recovery Project's website, welrp.org/maiinganthe-wolf-our-brother. As we will discuss in this chapter, its presence was most prominent at the height of the wolf debates, and its current absence might be due to its having served a timely purpose.

References

Adamson, Joni. 2001. *American Indian Literature, Environmental Justice, and Ecocriticism: The Middle Place.* Tucson: University of Arizona Press.

———. 2012. "Whale as Cosmos: MultiSpecies Ethnography and Contemporary Indigenous Cosmopolitics," *Revista Canaria de Estudios Ingleses.* 64 (1): 29–45.

———. 2014. "Cosmovisions: Environmental Justice, Transnational American Studies, and Indigenous Literature." In *The Oxford Handbook of Ecocriticism*, ed. Greg Garrard, 172–87. Oxford, UK: Oxford University Press.

Alia, Valerie. 2009. "Outlaws and Citizens: Indigenous People and the New Media Nation." *International Journal of Media & Cultural Politics* 5 (1/2): 39–54.

———. 2011. *Un/Covering the North: News, Media, and Aboriginal People*. Vancouver: University of British Columbia Press.

Allen, Chadwick. 2002. *Blood Narrative: Indigenous Identity in American Indian and Māori Literary and Activist Texts*. Durham, NC: Duke University Press.

Benton-Banai, Edward. 1988. *The Mishomis Book: The Voice of the Ojibway.* Hayward, WI: Indian Country Communications, Inc.

Borrows, John. 1997. "Living Between Water and Rocks: First Nations, Environmental Planning and Democracy." *University of Toronto Law Journal* 47 (4): 417–68.

Buffalo, Jean. 2007. "Reflections." In *From Dreams We May Learn (Paintings and Drawings,* illus. Rabbett Before Horses [Strickland], 21–47. Duluth, MN: Tweed Museum of Art.

Carroll, Clint. 2013. "Minnesota Wolf Policy Should Include Ojibwe Perspective." *Minneapolis Star Tribune*. March 13.

———. 2015. *Roots of Our Renewal: Ethnobotany and Cherokee Environmental Governance*. Minneapolis: University of Minnesota Press.

Clark, Bret. 2002. "The Indigenous Environmental Movement in the United States: Transcending Borders in Struggles against Mining, Manufacturing, and the Capitalist State." *Organization and Environment* 15 (4): 410–42.

"Concerns Over Possible Wolf Hunt in Minnesota." 2012. *KARE 11 News*. April 22. Accessed December 15, 2015. http://kare11.tv/1coBSvI.

"Declaration Designating the White Earth Reservation as a Ma'iingan (Wolf) Sanctuary." 2012. *Anishinaabeg News*. August 20. http://whiteearth.com/data/upfiles/files/wolf.pdf.

de la Cadena, Marisol. 2010. "Indigenous Cosmopolitics in the Andes: Conceptual Reflections beyond Politics." *Cultural Anthropology* 25 (2): 334–70.

Doerfler, Jill, Niigaanwewidam J. Sinclair, Heidi K. Stark, eds. 2013. *Centering Anishinaabeg Studies: Understanding the World through Stories*. Winnipeg: University of Manitoba Press.

Edberg, Jess. 2012. "Law Legalizes Wolf Hunting in Minnesota." *International Wolf Center*, June 13. Accessed 18 April 2016. http://www.wolf.org/wow/united-states/minnesota/erecoveryandmanagement/law-legalizes-wolf-hunting-in-minnesota/.

Eisenberg, Christina. 2010. *The Wolf's Tooth: Keystone Predators, Trophic Cascades, and Biodiversity.* Washington, DC: Island Press.

Erickson, Sue, and Jason Kekek Stark. 2012/2013. "Tribes Assert Sovereignty, Affirm Relationship with Ma'iingan." *Mazina'igan: A Chronicle of the Lake Superior Ojibwe* Winter: 1–2. http://www.glifwc.org/publications/mazinaigan/Mazinaigan.html.

Flesher, John. 2014. "Judge Orders Gray Wolves in Western Great Lakes Region Returned to Federal Endangered List." *Minneapolis Star Tribune*. December 14.

Fond du Lac Business Committee. 2012. "Fond du Lac Band of Lake Superior Chippewa Ordinance #07/12, Ma'iingan Protection." Resolution #1351/12. 31 October. http://www.fdlrez.com/government/ords/07-12ord.pdf.

Fritts, Steven H., Robert O. Stephenson, Robert D. Hayes, and Luigi Boitani. 2003. "Wolves and Humans." In *Wolves: Behavior, Ecology, and Conservation,* 2nd ed., ed. L. David Mech and Luigi Boitani, 289–316. Chicago, IL: University of Chicago Press.

Ginsburg, Faye. 2008. "Rethinking the Digital Age." In *The Media and Social Theory,* ed. David Hesmondhalgh and Jason Toynbee, 127–44. New York: Routledge.

Hallowell, Irving A. 1975. "Ojibwa Ontology, Behavior, and World View." In *Teachings from the American Earth: Indian Religion and Philosophy,* ed. D. Tedlock and B. Tedlock, 141–78. New York: Liveright Press.

Kahn, Richard, and Douglas Kellner. 2004. "New Media and Internet Activism: From the 'Battle of Seattle' to Blogging." *New Media & Society* 6 (1): 87–95.

Kraker, Dan. 2012. "Ojibwe Bands Ban Wolf Hunting—But Only on Indian-Controlled Lands," *Minnesota Public Radio*, October 31. Accessed December 15 2015. http://minnesota.publicradio.org.

Krech, Shepard. 1999. *The Ecological Indian: Myth and History.* New York: W. W. Norton.

LaDuke, Winona. 1999. *All Our Relations: Native Struggles for Land and Life.* Cambridge, MA: South End Press.

Latour, Bruno. 1993. *We Have Never Been Modern.* Cambridge, MA: Harvard University Press.

Loft, Steven. 2014. "Introduction to *Coded Territories: Tracing Indigenous Pathways in New Media Art*, ed. Steven Loft and Kerry Swanson, xv–xvii. Calgary, Alberta: University of Calgary Press.

Loft, Steven, and Kerry Swanson. eds. 2014. *Coded Territories: Tracing Indigenous Pathways in New Media Art.* Calgary, Alberta: University of Calgary Press.

Lynch, Tom, ed. 2005. *El Lobo: Readings on the Mexican Gray Wolf.* Salt Lake City: University of Utah Press.

maiingan.org n.d. (Website). Accessed 16 January 2015. [Website no longer available.].

Martin, Kathleen J. 2012. "Traditional Responsibility and Spiritual Relatives: Protection of Indigenous Rights to Land and Sacred Places." In *Indigenous Rights in the Age of the UN Declaration*, ed. Elvira Pulitano, 198–227. Cambridge, UK: Cambridge University Press.

Mech, L. David, and Luigi Boitani. eds. 2003. *Wolves: Behavior, Ecology, and Conservation,* 2nd ed. Chicago, IL: University of Chicago Press.

Minnesota Department of Natural Resources. 2001. "Minnesota Wolf Management Plan." Accessed December 15, 2015. http://files.dnr.state.mn.us/natural_resources/animals/mammals/wolves/wolfplan.pdf.

Nadasdy, Paul. 2005. "Transcending the Debate Over the Ecologically Noble Indian: Indigenous Peoples and Environmentalism," *Ethnohistory* 52 (2): 291–331.

———. 2011. "'We Don't Harvest Animals We Kill Them': Agricultural Metaphors and the Politics of Wildlife Management in the Yukon." In *Knowing Nature: Conversations at the Intersection of Political Ecology and Science Studies*, ed. Mara J. Goldman, Paul Nadasdy, and Matthew D. Turner, 137–41. Chicago, IL: University of Chicago Press.

Niezen, Ronald. 2003. *The Origins of Indigenism: Human Rights and the Politics of Identity.* Berkeley: University of California Press.

Pavlik, Steve. 1999. "San Carlos and White Mountain Apache Attitudes toward the Reintroduction of the Mexican Wolf to Its Historic Range in the American Southwest." *Wicazo Sa Review* 14 (1): 129–45.

Pember, Mary Annette. 2012a. "Cry, Wolf: The Decision by the State of Wisconsin to Kill Wolves for Sport Has Outraged Local Tribes." *This Week from Indian Country Today,* September 5: 22–25.

———. 2012b. "Wisconsin Tribes Struggle to Save Their Brothers the Wolves from Sanctioned Hunt." *This Week from Indian Country Today,* August 14.

Pulido, Laura. 1998. "Ecological Legitimacy and Cultural Essentialism: Hispano Grazing in Northern New Mexico." In *Chicano Culture, Ecology, Politics: Subversive Kin,* ed. Devon G. Peña, 121–40. Tucson: University of Arizona Press.

Ranco, Darren. 2007. "The Ecological Indian and the Politics of Representation: Critiquing The Ecological Indian in the Age of Ecocide." In *Native Americans and the Environment: Perspectives on the Ecological Indian*, ed. M. E. Harkin and D. R. Lewis, 32–51. Lincoln: University of Nebraska Press.

Robyn, Linda. 2002. "Indigenous Knowledge and Technology: Creating Environmental Justice in the Twenty-First Century." *The American Indian Quarterly* 26 (2): 198–220.

Ross, Anne, Kathleen Pickering Sherman, Jeffrey G. Snodgrass, Henry D. Delcore, and Richard Sherman. 2011. *Indigenous Peoples and the Collaborative Stewardship of Nature: Knowledge Binds and Institutional Conflicts.* Walnut Creek, CA: Left Coast Press.

Smithers, Gregory D. 2015. "Beyond the Ecological Indian: Environmental Politics and Traditional Ecological Knowledge in Modern North America." *Environmental History* 20 (1): 83–111.

Srinivasan, Ramesh. 2006. "Indigenous, Ethnic and Cultural Articulations of New Media." *International Journal of Cultural Studies* 9 (4): 497–518.

Treuer, David. 2008. "Seeing Red: A Portrait of the Artist." In *From Dreams We May Learn (Paintings and Drawings),* illus. Rabbett Before Horses Strickland, 11–20. Duluth, MN: Tweed Museum of Art.

Tsosie, Rebecca. 2007. "Indigenous People and Environmental Justice: The Impact of Climate Change." *University of Colorado Law Review* 78: 1625–1677.

Whyte, Kyle Powys. 2013. "Justice Forward: Tribes, Climate Adaptation and Responsibility." *Climatic Change* 120 (3): 517–30.

Willow, Anna J. 2009. "Clear-Cutting and Colonialism: The Ethnopolitical Dynamics of Indigenous Environmental Activism in Northwestern Ontario." *Ethnohistory* 56 (1): 35–67.

Wolfe, Patrick. 1999. *Settler Colonialism and the Transformation of Anthropology: The Politics and Poetics of an Ethnographic Event*, New York: Cassell Publishing.

Wolfley, Jeannette. 1999. "Ecological Risk Assessment and Management: Their Failure to Value Indigenous Traditional Ecological Knowledge and Protect Tribal Homelands." In *Contemporary Native American Cultural Issues,* ed. Duane Champagne, 293–306. Walnut Creek, CA: Alta Mira Press.

7 Cyclical Conceptualizations of Time

Ecocritical Perspectives on Sámi Film Culture

Pietari Kääpä

The Sámi people, a minority estimated at 164,000 individuals globally, primarily inhabit the Sápmi, a contested geocultural space covering northernmost Norway, Sweden, Finland, and Russia. The complex modes of colonial governance over this space have led to its being described as either transnational (Helander and Kailo 1998) or as postnational (Kääpä 2014), The Sámi have moved over time from being outsiders within the European nation-state system to being recognized as a national "minority," and from there to becoming an acknowledged (if "protected") part of egalitarian society. Because of the long historical contestations that mark these transitions, the politics of Sámi Indigeneity are frequently discussed as part of the representational modus operandi of Sámi culture. Sámi cinema is an integral part of this representational culture and has gained attention in the emerging Indigenous cinema study movement. (See, for example, Columpar 2010, Dahlqvist 2014, Ginsburg 2011, Mecsei 2014, and Raheja 2007). The history of Sámi cinema also features heavily in recent work on the cinema of the Arctic (MacKenzie and Stenport 2014), which, alongside discussions with Sámi filmmakers and participation at Sámi film festivals (such as the Skabmagovat), attests to the idea that, for the Sámi, cinema is an important means of self-representation and recognition.

I have addressed these perspectives in my earlier work (Kääpä 2014), but in this chapter I want to articulate an alternative perspective on the politics of recognition by investigating ecocritical considerations in Sámi film culture. These ecological themes, I argue, speak to the flexibility and complexity inherent to Indigenous cosmopolitics, a key concept explored in this volume. Below, I address this claim by first, describing how cinema plays a part in sociopolitical debates about Sámi and non-Sámi "ecological temporality" and by connecting such temporal understandings to the idea of cosmopolitics; second, by relating the relevance of such understandings to perspectives on sustainability; and third, by highlighting Sámi cinema's distinct notion of "banal everydayness." As we shall see, Sámi cinema helps its audience visualize alternative ecological existential modes to hegemonic notions of progress and stereotypical ideas of Sámi as the exotic other, even as it testifies to diversity and complexity within Sámi identity.

Linear and Cyclic Time and Indigenous Cosmopolitics

The Sámi, at least in more traditional conceptions, base their perspectives and ideas of *dasein,* of being in the world, on a distinctly cyclical approach to time and temporality. Cyclicality emphasizes the recurring and holistic nature of existence, both in terms of natural resources and the lifecycle of humans. Knox (2001) uses the words of respected Sámi author Nils-Aslak Valkeapää to evoke this central notion of cyclicality:

> I lived
> lived a complete life
> one day followed another
> rose sank disappeared arrived
>
> (Knox 2001, n.p.)

The words of the poem capture the sense of cyclicality inherent to natural lifecycles and seasons. Knox argues that the words capture the life structure of reindeer herders who take their herds to the mountains in the summer and move into the forests for the autumn. The point of these movements is not to increase the herd or its value as a source of capital but instead to allow the reindeer to follow their natural rhythm. While seasonal variations will impact herding patterns each year, there is a considered logic to the practice; a system in place that sustains and preserves the land from year to year. Such cyclicality is considered integral to the Sámis' self-conceptualization of their place in the world.

In contrast to such cyclical perspectives, hegemonic cultures of the region emphasize linear conceptions of time where renewability is not a priority so much as ensuring that renewable sources are maximized. A linear mode of time is anthropocentric, primarily seeing value in increasing human production and consumption. In dominant Nordic cinema, which privileges linear time, Sámi cyclicality, or ways of being in the world, have rarely been acknowledged in any complex depth. Instead, in Norway and Finland, the two countries with the most active and contested histories of Sámi film production, film culture has relied on stereotypes of an exotic, mysticized other. These stereotypes have been discussed with some frequency in academic literature (Kääpä 2014; Soila, Widding, and Iverson 1998), especially in Finland, where Finnish producers have depicted the Sámi as either villains or mystical forces of nature in highly successful films such as *The Curse of the Witch* (1927) and *The White Reindeer* (1952), respectively. In both these cases, the focus is on conceptualizing northern "Lapland" as an untamed territory and wilderness space that needs to be civilized. Such processes of civilization usurp and colonize both the people who inhabit these spaces as well as their natural resources. This subjugation is part of a long history. For example, the Finnish national epic *Kalevala*, a collection of folk poems first organized into narrative form by Elias Lönnrot, a member of the Finnish cultural elite in the mid–nineteen century, has such colonial tendencies. Literary critic Pentikäinen (1989) addresses the complex forms of colonial symbology in this text, while the ethnologist Kailo (2001) highlights problematic relations between gender and subjugation.

In most non-Indigenous Scandinavian cinema, the Sámi characters act predominantly as symbols of difference, often used to evoke threats to the settler's order. The role they occupy in relation to the land is dominated by simplistic assertions of their animistic qualities (as with the protagonist of the 1952 film, *The White Reindeer,* who transforms into a homicidal reindeer) or their nomadic nature. Both of these qualities suggest the Sámi are unreliable and corrosive suitors for women from the dominant population (see, e.g., *The Earth Is a Sinful Song,* 1973). *The White Reindeer* draws a parallel between the Sámi and the reindeer, which are portrayed literally as transformative animals, while in *The Earth Is a Sinful Song*, the Sámi are described as laborers, only suited to the simple task of herding. In contrast, the majority of Finns are depicted as capable and stable forces of civilization.

While popular, non-Indigenous depictions of the Sámi have since become more nuanced, some of these stereotypical representations persist even in films with distinctly egalitarian takes on northern communities. Mainstream productions such as *Umur* (2002) and *Wolf* (2008) feature protagonists who are not explicitly coded as Sámi, but who have been identified as such by the publicity surrounding the films. In these films, the characters lack the cultural signifiers—ethnic dress, shamanistic customs—emphasized in earlier cinema. Yet, both films frame their Sámi characters in restrictive terms (see Kääpä 2014). For example, the protagonist of *Wolf* is a reindeer herder whose attempts to protect his herd place him in conflict with the Swedish state over animal protection rights. In *Umur*, the object of affection for the film's male protagonist, a character called Umur, is caricatured as a mythic force of nature, a creature that seems to exist beyond the protagonist's understanding. Both "modern" Sámi protagonists are individuals unable to free themselves from their natural connections and who are either imprisoned or die due to this connection. In such narratives, the Sámis' environmental immersion acts as the ultimate tool of othering.

Elsewhere, I have addressed these considerations in the framework of the politics of "giving and withholding recognition" (Taylor 1997), whereby any sense of recognition becomes a political tool of control. While it would be problematic to disavow Sámi cultural politics from being part of the multiculturalist "permissiveness" of the neoliberal state, the "naturalization" of Sámi culture, including the role of cyclicality, enacts what Rob Nixon has called slow violence (2001). This is a form of violence initiated on the world's poor—including many Indigenous peoples—via environmental disasters (loggings, deforestation, climate change, toxicity, etc.) that are deemed too slow, and thus too inconsequential, in their harm to warrant political or ethical attention. A crucially significant aspect of slow violence is the persistent but often unacknowledged representational erasure of Indigenous cultures into hegemonic rationality. The Sámi are granted access to the constitution of a hegemonic state, but only as "others," who do not fit in with normative ideas of linear progress. Representational erasure and othering in mainstream cinema accentuates such slow violence, inculcating a sidelining of Sámi cultural ways of knowing, and prioritizing, as I discuss more fully below, a linear mode of time that values capital resourcism, not cyclical renewal.

Scholars like Helander and Kailo argue that "the Sámi are, of course, a modern people who must live according to linear time. Nevertheless, the references in Sámi scholarship to the value of a cyclical model ordering life are too numerous and central to be ignored" (1998, 208). Thus, evoking key areas of Sámi culture in contemporary society should take into account these complexities. The emergence of a dialogic perspective that facilitates multiple levels of comprehension ties in with what Joni Adamson suggests is awareness of "nature's multiple and heterogeneous ontologies … [where] multiple possibilities might weigh into discussions of differing, even competing political proposals without any of these proposals being dismissed as 'right,' 'left,' 'superstitious' or 'unscientific'" (Adamson 2012, 159). Such flexible and dialogic understandings carve out a more complex communication space for exhibiting Sámi experience and can be understood via the idea of Indigenous ecologies, or the more frequently used term, "cosmopolitics."

Drawing on the work of Bruno Latour and Isabelle Stengers, Marisol de la Cadena (2010) characterizes cosmopolitics as a means of unraveling the pervasive hegemonic distinction between modernity and nature. Latour's perspective on "the cosmos" is an inclusive one, incorporating human and nonhuman beings and objects as equal participants in constructions of meaning to challenge the ontological certainties of Western life-worlds. Stengers' work on cosmopolitics elaborates on these ideas, suggesting that a truly egalitarian politics can be ascertained by building "worlds alongside—if not necessarily with—those working from wholly different blueprints" (Stengers 2005, 995). de la Cadena, in turn, uses this approach to consolidate a space for Indigenous self-conceptualizations that seek to redefine the idea of politics. For her, cosmopolitics "aims at transforming the concept from one that conceives politics as power disputes within a singular world, to another one that includes the possibility of adversarial relations among worlds: a *pluriversal* politics. The first step is to recognize that the world is more than one socionatural formation; the second is to interconnect such plurality without making the diverse worlds commensurable" (de la Cadena 2010, 360–61). de la Cadena uses the term *equivocations*, meaning "more than one but less than two," which is an approach that refuses the universalizing connotations of oneness but also identifies the problems of a binary division into two distinct spheres.

The concepts of pluriversalism and equivocations are particularly useful, as they indicate the complex subjective ecologies that constitute Sámi calls for self-recognition and diversity. This is especially the case with the role of Sámi film production within hegemonic film industries. In contrast to the slow violence perpetrated by hegemonic cinema, much of contemporary Sámi cinema seeks to balance traditional and "modern" cultures. In doing so, such cinema attempts to find both a space for vocalizing Sámi sovereignty alongside hegemonic political projects such as the United Nations Declaration on the Rights of Indigenous Peoples and a vernacular that incorporates the philosophy of "more than one but less than two" into cinema.

To evaluate how cosmopolitical considerations emerge in contemporary Sámi cinema, I initially turn to films produced by non-Sámi directors such as Hannu Hyvönen. These films illustrate some of the political concerns that structure Sámi struggles for sovereignty, such as land ethics and resource politics. They also indicate some of the prescient ways in which Sámi directors express such themes differently.

The Politics of Indigeneity and Sustainability

Hyvönen's film *Last Joik in Sámiland* ("*Viimeinen joika Saamenmaalla*" 2009) provides a powerful argument for understanding the over 10,000-year ecological complexity of the Sápmi. Much of this facilitation takes place through a particular vision of sustainability that revolves around the herding patterns of reindeer. The film explains how reindeer herding is adjusted to cyclicality as the herds find their sustenance from the forests and depend on natural rhythms. The reindeer rely on lichen and moss that grow on trees of a certain age, but the more recent and persistent harvesting of trees by large corporations such as Stora Enso (a public-private company partially owned by the Finnish state), in collaboration with organizations like Metsäliitto (a cooperative organization focused on the foresting business), has exposed the soil and destabilized these complex forest ecosystems.

The opposition of cyclical and linear life perspectives becomes the key focal point of *Last Yoik in Sámiland* (2008). An ecologist explains the rationale for protecting the forests, emphasizing that sustainable foresting is not only in the best interests of the Sámi but also of the companies as well. One of the environmental representatives for the company argues that only sixteen percent of the total area of the Sapmi is forested, but the film counteracts this by suggesting that the company does not take into account the impact of logging on the forests' long-term ecology. The reindeer live off lichen, which can survive if there is sufficient cover for it (the trees need to be at least three meters high), but lichen and consequently reindeer will only truly thrive in what local activists call a "hundred-year" forest. As this obviously necessitates an ecological cycle lasting a hundred years, it is no surprise that the reindeer are in danger. "Responsible logging" decimates specific defined areas, as the only focus is on the parts versus the whole; the forests consist of young areas that dry up due to exposure to too much sunlight or wind, which makes sustaining the Sámi ecological system impossible. As reindeer are not able to continue to graze the land when trees disappear, the reindeer population will eventually wither away. The process in turn leads to a destabilization of the Sámi lifestyle and results in battles for rights and livelihood, as we see in this film.

The majority of *Last Yoik in Sámiland* shows discussions with locals who are concerned with the immensely destructive impact of these logging practices. In contrast to the perspective of the Sámi, interviews with corporate spokesmen emphasize the scope of the operations of Metsäliitto and Stora Enso. The two perspectives could not be more different. This is a division that reflects not only the linear/cyclical distinction but also a fundamental power imbalance. A Sámi local explains the ways his small-scale herd is not able to return to its traditional place next year. In contrast, one of the managers of the company discusses the ways they harvest and replant trees to maximize the cycle in which they grow and can be used for product. The foresting companies ignore a key ecological concept premised on the agency of these areas and animals as active nonhuman actors that contribute to the long-term sustainability of the region.

The complicity of Metsäliitto with state power is, at least from the perspective of the makers of *Last Yoik in Sámiland*, one of the fundamental reasons for a continued imbalance in Sámi-Finnish power relations. At stake here are the different perspectives of sustainable Sámi herding and the hegemonic ideology of growth, whose logic of sustainable development emphasizes the latter word in the term (see, e.g., Finnish Ministry of the Environment 2011).[1] To exemplify how these ideas lay the foundations for the film's arguments concerning approaches to environmental resources, the film presents both Sámi and company perspectives. "Environmental managers" for Stora Enso are especially proud of the fact that they employ 45,000 people in various roles and operate on five continents. In contrast, Indigenous activists such as Nilas Somby explain that while the Sámi do have their own water and land rights, they are not met with any level of respect by the large companies partially owned or working in collaboration with state organizations. Thus, the film makes it explicit that Indigenous rights are integrated into corporate logics only in so far as they comply with resource consumption and production.

The division of these perspectives avoids any pluriversal understanding, a fact that the Sámi Council wants to undo. As shown in the film, they appeal to the United Nations Council and Greenpeace but, as a representative points out, the party that has the power can set their definitions of what qualifies as correct modes of behavior. Thus, the representatives express a lack of faith in the negotiations, which ties in with subsequent reversals to the aspirational connotation of the United Nations Declaration on the Rights of Indigenous Peoples (UNDRIP). *Last Yoik in Sámiland* suggests a range of reasons complicating the successful adaptation of the UNDRIP, mostly concerning private and public interests, which tend to override Indigenous causes. Metsäliitto is one such party, as it has been behind a variety of campaigns to demean Sámi ownership of the land and forests. As the film demonstrates, newspapers have waged a campaign to undermine Sámi ownership, with the Northern Finnish newspaper *Inarilainen* publishing a statement that the state may not be able to give the Sámi rights to the land because they never had such rights in the first place. To make matters worse, the film shows that these arguments soon escalate into aggression, as several Greenpeace activists as well as members of the Sámi council are threatened with physical violence by a group of anti-environmentalist protestors. The Sámi are shown lodging an official complaint with the United Nations Council for Indigenous Peoples in 2005 and managing to halt the logging. Yet legislation on the state level allows the companies to move to adjacent regions and practice unsustainable use of forests there with few limitations.

Ultimately, *Last Yoik in Sámiland* concludes in a pessimistic tone as it makes clear that these are issues of global capitalism, which in this context is presented as fundamentally opposed to cyclical life perspectives. The film foregrounds both Sámi traditions and the central role cyclicality holds for them. It culminates with an impassioned plea for returning the land to the Sámi but also acknowledges that this is unlikely. The film does this in a matter-of-fact style, which complements its attempts to increase awareness of Sámi exploitation and draw attention to sustainability rhetoric in official policy and business documents. For example, *Environmental, Energy and Natural Resources Research in Northern Finland*, a document compiled by the Northern Research Information Platform and the University of Oulu (Suorsa and Iskanius 2010), provides a comprehensive overview of sustainable development in the area. It chronicles a range of mechanisms used by public and private organizations that utilize the "riches" available in Northern Finland, as well as mechanisms to protect the sustainability of the resources. Much of the report emphasizes the vernacular of growth, including sections on minerals and forestry. Reindeer herding is mentioned twice in the whole publication as a potential use of natural capital and as a protectable means of sustenance.

The rhetoric here clearly prioritizes linear progress, marginalizing any notion of Sámi cyclicality. In contrast, a recent document on sustainability published by the Ministry of the Environment (2015) proposes a domestic social agreement that lists the following areas as key concerns: (1) equal access to welfare, (2) a society of citizens with means to influence, (3) sustainable work, (4) sustainable residencies and local communities, (5) a carbon neutral society, (6) an economy mindful of resources, (7) lifestyles respectful of nature's means, and (8) decision making respectful of nature (Finnish Ministry of the Environment 2015). The majority of these designators for future policy developments highlight similar ideas proposed by the Sámi in *Last Yoik in Sámiland* with an additional emphasis on human rights. Whereas earlier state policies enforce linearity, even on the level of state legislation, this document gestures to a more productive understanding of sustainability. While operating on the level of reportage, *Last Yoik in Sámiland* emphasizes the problems traditional Sámi cyclicality and its related life philosophies face in the contemporary world. Simultaneously, the pervasive sense of anger people in the film exhibit reminds us that the Ministry of the Environment's commitments to Sámi perspectives in sustainable development encourages only voluntary participation, a practice that brings into question any potential for challenging the marginalization the Sámi. This serves to remind us that many of the policy examples that speak to cosmovisions are also often relegated to aspirational goals and do not become a set of concrete regulations to follow.

Politics of Sovereignty

The politicized argument of *Last Yoik in Sámiland* aims to establish space for consolidating pluriversal understandings of sustainability. Yet, its rhetoric necessarily adopts an antagonistic tone that follows an oppositional binary logic instead of the type of hybridity necessary for pluriversalist politics. While films by Sámi directors also work with similar distinctions between hegemonic and Indigenous cultures, they have often sought to blur distinctions between the norms.

Perhaps the best known films are Nils Gaup's *Pathfinder* (1987) and *The Kautokeino Rebellion* (2008). Distributed widely with considerable critical acclaim internationally, both have been discussed in scholarly sources, both as Norwegian genre films (Soila, Widding, and Iverson 1998) and as texts specifically focused on minority politics in Sámi culture, (Nestingen 2008). Both use conventional narrative structures (the former of the Western, the latter of the historical epic) in constructing re-creations of the politics of sovereignty, but they combine these structural "concessions" with a range of elements Indigenous to Sámi culture. Below, I focus on *Pathfinder* because of its seminal place in Sámi film history (including a nomination for Best Foreign Picture Oscar) and its representation of Sámi mysticism, which evokes ecological cyclicality as a manifestation of pluriversality.

Gaup's "Northern" (a play on the term "Western," first coined by Gunnar Iversen, [1998]), set in unspecified precolonial times, describes the clash of two opposing lifestyles. Aigin, the protagonist, who is coded as Sámi, loses his entire extended family when they are slaughtered by an invading group, The Chude. We are provided very little information about the Chude outside of their colonialist tendencies, which makes them akin to contemporaneous colonialist powers with whom the Sámi have had to contend throughout their history. This outside ethnic group, all clad in sleek black leather and working at a level of military efficiency, slaughters Aigin's people and head next to a neighboring village. However, Aigin is able to warn these villagers and a majority leave for the coast. The remaining men are no match for the marauders who kill them. Aigin is the only one left alive. Before he is captured he has a vision of a shaman who prepares him for the tasks ahead. The shaman tells Aigin that humans are part of an ongoing bond with the cosmos that cannot be broken. The concept of the eternal bond is a clearly cyclical evocation that binds both Aigin and the whole of the Sámi to the environment that surrounds them. In contrast, the Chude, bound by hate and thirst for power, are not able to see or feel these cosmological bonds. It is not difficult to see them as embodiments of a linear life philosophy where accumulation and consumption overtake any notion of sustainability or harmony with nature. The film makes this contrast between the two worldviews explicit with multiple symbolisms. Early on, we see a hunting ritual where the villagers honor a captured bear. The procession includes a ceremony to ensure that the sacrifice of the bear is acknowledged by the villagers. In contrast, the Chude raid Aigin's farm and simply steal what they want and kill the rest of his family. The Chude see life as a thing to consume and possess.

The events chronicled in the rest of the film comprise a rite of passage for Aigin as well as a way to cinematically gesture towards a cyclical universe. Initially, Aigin, blinded by vengeance, is unable to see the cyclic bond. The Chude force Aigin to lead them on a precarious route through the mountains in their pursuit of the coastal Sámi. The rope that binds Aigin to the Chude (literally and symbolically) ultimately serves to clarify their different life perspectives. The Chude have symbolically misappropriated the bonds of a cyclical universe as a tool of destruction and tie Aigin to their destructive worldview. For Aigin the rope is a means to save his people—during a particularly treacherous path, Aigin cuts the rope, sending the Chude falling to their deaths. He severs his own ties to their worldviews of greed as he finally is able to comprehend the complex patterns of interconnectivity that underline the Sámi life philosophy. Much as in the earlier discussion of political uses of sustainable development, this symbolic idea outlines another evocation of how key ecological meanings can be perceived differently. By traversing between the Chude and Sámi perceptions of connectivity, the film outlines a pluriversal approach to the world, or of one's place in the cosmos, where the Chude, unlike the Sámi, act not as a clearly identifiable ethnic tribe, but as a metaphoric representation of an enclosed worldview.

Pluriversalism also emerges in the films of Paul Simma, a prolific producer of Sámi cinema since 1991. Simma's productions range from fiction films such as *The Minister of Sagojoga* (1997) to documentaries including *Give Us Back Our Skeletons!* (1999). I have discussed these films in earlier work (Kääpä 2014). Here, I focus on the short film, *The Tale of Arctic Love* (2000), which combines explorations of contemporary Sámi life with surreal forms of mysticism. The main protagonist of the film, Iisko-Matti, has agreed to meet the daughter of a local herder at her grandmother's funeral. His job as a mechanic complicates things, as a rich Swedish patron demands his snowmobile be fixed on the same date. The conflict is largely pitched as one between loyalty to culture and the lures of financial success. Attempting to reconcile the two, Iisko-Matti takes the snowmobile to the burial ceremony and is subsequently requested to transport the dead woman's corpse on the snowmobile to her resting place.

The journey acts as a convenient detour from the expected linear narrative. Instead of working on its romantic comedy plotline, the film veers into the mystical spirit world of the northern regions. To emphasize this alternative life-world, the dead woman wakes up mid-journey and scares Iisko-Matti up a tree. While this is a convention that seems to adhere to the horror film, in *The Tale of Arctic Love* it acts as a rite of passage, as the next day, Iisko-Matti is commended for his bravery for having ventured into the otherworld. The collision between a modern lifestyle and mysticism leads Iisko-Matti back to traditional Sámi culture. His role until this moment suggests that of an outsider, as he mostly communicates in Finnish and is not considered welcome at the funeral. Now, having demonstrated his link to the spirit world, Iisko-Matti is provided with enough money by the community to compensate for the broken snowmobile and is even allowed to ask for the daughter's hand in marriage. It seems that even in the hegemonic society ruled by capital, a connection with a parallel Indigenous life-world can provide the means to enter a more plural understanding of one's position in society.

Importantly, the inclusion of mysticism in *The Tale of Arctic Love* and *The Pathfinder* complicates stereotypical exoticized portrayals of Sámi, particularly those of the "ecological Indian"—a contested concept encapsulating a history of othering of Indigenous peoples premised on essentialist naturalization (see Columpar, 2010; Harkin and Lewis 2007; Krech 1998; Monani 2014). Instead of othering, the mystical angle works inclusively, blending contemporary life styles and means of representation with traditional worldviews. For this, the films evoke Sámi worlds in various ways, from the symbolic power of rituals honoring slain animals to shamanistic shape-shifting practices, and make room for perspectives alongside modernity's narrative of environmental progress. *The Tale of Arctic Love's* combination of the range of references associated with dominant cultures (snowmobiles, cars, mainstream entertainment, and endless possiblilities for travel) and the "unexplained" comprises a direct challenge to the singular logic of hegemonic domination. Unlike the oppositional connotations of mainstream film culture—whether of the ignoble primitive from Nordic films like *The White Reindeer* or of the noble but soon-to-die ecological Indian in *Last Yoik in Samiland* (one possible interpretation)—Simma and Gaup's films blur binary distinctions of worlds and worldviews and gesture to the "more than one, but" approach of pluriversal cosmopolitics.

Banal Everydayness

Pluriversal evocations in Simma and Gaup's films are based on asserting the uniqueness of Sámi culture. However, such uniqueness has also been interpreted as something of a catch-22 by scholars such as Kailo (2001). To highlight such catch-22 situations in Indigenous film, scholars such as Columpar (2010), Kääpä (2014), and Monani (2014) all draw on

Gayatri Chakravorty Spivak's concept of strategic essentialism (Spivak 1988), which points to how the representation of traditional worldviews are, on the one hand, useful to the politics of recognition, but on the other hand, because they are distinct from hegemonic culture, they can be interpreted in ways that other and further marginalize Indigenous people. As Columpar (2010) and Monani (2014) suggest, Indigenous filmmakers will turn to "anti-strategic essentialism" to counter interpretative possibilities of stereotypical marginalization. Here, I link the idea of anti-strategic essentialism to what I call "banal everydayness," a mode of seemingly depoliticized representation that concerns the politics of performing the everyday. The term is a variation on Michael Billig's (1995) concept of banal nationalism, which describes a range of means of evoking nationhood without engaging in explicit nationalist rhetoric. He talks of the often unnoticed presence of national flags as a way of reminding citizens about nationhood without evoking the nation through explicit visual means. In terms of Sámi politics it is not too difficult to identify representations that show Sámi life through more complex (or alternatively more deglamorized) means than a traditional set of strategic representations. The idea here is that the "Sáminess" of a film stays in the background while it, obviously, maintains a considerable level of significance for the film's thematic scope.

Banal everydayness attempts to capture Sámi contemporary situations that, as Mecsei describes, are characterized by hybridity as "the majority of the Sámi population do not have an evident connection to the images that have been, and still are, put forward as traditional Sámi culture or territory" (2014, 79). While it would be counterproductive to dismiss the cyclical adventures of *The Pathfinder* or the sustainability arguments of *Last Yoik in Sámiland*, it is also clear that reindeer herding and notions of cyclicality may not meet the reality of most contemporary Sámi life.

Cry into the Wind (2007), a documentary film by Koltta Sámi Katja Gauriloff focuses on the ordinary everyday practices of individuals living in the Sápmi. By showing us the travails of being a municipal politician or a teenager in a small northern community instead of focusing on being in a Sámi community, the film facilitates a cinematic evocation of diversity within the Sámi, or of an internalized version of pluriversal cosmopolitics. The politics of the film are not so much about establishing sovereignty, but about encountering geographical marginalization, reflected in the choice of the film to focus on the day to day administration of Sámi communities as well as the meager opportunities the small scale and isolation have on the youth of these communities. Any sense of cyclicality, in a more obvious sense, is missing.

The approach of the film echoes what Bang and Sorenson calls new forms of political participation where citizens act as "everyday-makers" (1998, 327–329). Instead of an enthusiastic performance of civic or national identity, it is the mundane tasks that evoke a more accurate picture of citizenship as an act of multiple belonging, or of evoking complex Indigenous cosmopolitics.[2] The act of representing such everydayness in *Cry in the Wind* produces an ethnic variant of banal nationalism by evoking the Sámi community without foregrounding that community. In terms of Sámi iconography, the occasional presence of traditional costumes or the flag of the Sámi on a table remind viewers that stories about daily life in a northern Finnish community are really about demonstrating the contemporary existence of the Sámi in and as part of the Finnish nation, even as they are subjugated to the norms of that context. These banal evocations of performing Sáminess emphasize both the constant need to administer Sámi identity and the ways the activities they perform at places like schools or municipalities confirms marginality and efforts for recognition. What makes these particular depictions significant is that the banal divests these cultural negotiations of the sort of exoticness perpetuated in hegemonic representations through imaginaries of the mythical Sámi. Such perspectives are vital in evoking a more complex form of Sámi identity beyond the simplistic othering of exotic depictions of the reindeer herding mythical Sámi. But simultaneously, it is easy to see how critics may perceive the "the private, the ordinary, the everyday" (Turner 2010, 22) as another set of hegemonic strategies whereby "citizens are voluntarily co-opted in projects of power that may have a therapeutic value in 'giving voice' to the ordinary but ultimately reproduce local asymmetries and global inequalities" (Chouliaraki 2010, 230). To put it simply, by showing the Sámi as a minority whose goals and wishes are similar to "everyone," they will not only be considered as a minority by audiences outside of their community, but also as one that shares the hegemonic population's qualities of normalcy and worth.[3]

Gauriloff states that she encountered some criticism for depicting contemporary Sámi life too negatively, but argues it is precisely her aspiration to show people not only as Sámi but also as individuals (Gauriloff 2012). Sámi understandings of cyclicality and sustainability do not make explicit appearances here. Indeed, it seems more exotic or mythical associations need to be disavowed in order for the film's politics to make sense. Is this a concession to hegemonic norms or a particularly effective case of anti-strategic essentialism? Suggesting that the film goes for the former would be misunderstanding its intentions. Instead we can consider it as a case of what Monani (2013) has labeled oblique eco-testimony, where "eco-identity permeates Indigenous experience outside battles for environmental conservation and autonomy." The combination of political content with the seeming banality of its everyday perspective situates Gauriloff's film as an instance of cinematic pluriversalism. This is another form of cosmopolitics, albeit one where the emergence of a plurality of life-worlds takes place within Sámi culture. After all, to be able to argue for a comprehensive cosmopolitics, one must also account for differences within one's own culture.

Producing Cyclicality

The history of Sámi cinema could be seen as a narrative in which the Sámi emerge from outsiders to a minority and from there to an acknowledged, if "protected" part of an egalitarian society. Such a narrative is, of course, problematic in two key ways. Firstly, the role it affords the Sámi objectifies them as auxiliary to mainstream historical progression. Secondly, it does this in the language of accumulation and progress, which stands in contrast to Sámi cultural and societal norms. Through a series of textual analyses I have suggested above that a more balanced approach to Sámi cinema comes from considering its pluriversal politics. Below, I suggest that such pluriversal understandings are also important when considering the industrial concerns of Sámi film production culture.

The success of films like *The White Reindeer* have contributed to Sámi marginalization whereas *The Pathfinder* has challenged such perspectives. But simultaneously, these productions have benefitted from large budgets and distribution networks and have in one way or another also contributed to public perceptions of the Sámi, mostly negative in the former, positive in the latter. The works of Simma and Gauriloff are marginalized, as they are only available on very limited VHS or DVD releases or, as in the case of *Cry in the Wind*, which only received two showings on the national television channel, YLE2, the second one of these at 11:25 A.M. on a weekday. *Last Yoik in Sámiland* has been screened at festivals but is now available only on the internet. The treatment of the Sámi protestors captured in *Last Yoik* would have made for politically intriguing material for Simma as well, "but YLE was not interested, it was impossible to do financing independently and the foresting industry is too powerful" (Simma 2015). For Gauriloff, too, official support was lacking: "With *Cry into the Wind*, the Finnish Film Institute did not want to fund it, as they had apparently done a Sámi project a few years back. It was not for AVEK (The Centre for the Promotion of the Audiovisual Arts), even though they did provide script-writing assistance. National Geographic gave us a grant for shooting the film. I think the topic was not interesting or sexy enough for FFF [the Finnish Film Foundation] or AVEK" (Gauriloff 2012).

Clearly, any sense of "sexiness" in Sámi film would concern exotic myopia of Finnish films or the epic historical scope of Gaup's films. Another complication comes from the difficulty of representing cyclicality on screen. Paul Simma suggests (2015) that the oral tradition, in particular, poses obstacles as the traditional, often cyclic, modes of storytelling are difficult to transcribe into competitive scripts. As Sámi producers have to work in small nation film industries, they have to convey their ideas to cultural authorities such as the Ministry of Education and Culture in Finland as well as the Finnish Film Institute. Constructing cyclicality in narratives is a problem as the Sámi film producers do not enjoy protected status as a minority despite occasional earmarked funds to bolster infrastructure (Kääpä 2014). In my interview with him, Simma (2015) explains the complications of producing these films in the contemporary industry:

> Ten years ago it was relatively easy to produce films in the Sámi language as one could automatically receive distribution in both Finland and Sweden. Today it is more difficult as infrastructural support for minority productions does not exist on a stable level. If one wants to produce Sámi films they have to be made for television. The numbers for audiences viewing these films will be comparatively high but such productions will fail in cinemas.

Indeed, the most considerable challenge would seem to come from the conventions of cinematic form whereby audience expectations often favor eventfulness and specific plot structures with neat conclusions. While art film has challenged such conventions at worldwide festivals with the "unconventional" films by a range of acclaimed directors from Abbas Kiarostami to Jia Zhang-Ke, Sámi filmmakers are even further marginalized from art house productions because of the smaller scale of their production and distribution arrangements, yet they have to work in the context of film industries that prioritize classical narrative structure: "I struggle with how to transfer these stories that are so different in form to those with a clear beginning and end. It is a real difference in storytelling and it is not easy to combine the two" (Simma 2015).

The politics of producing Sámi cinema are also complicated by funding issues. When these considerations feature in domestic politics, they adhere to what could be called "permissive tolerance" or a superficial acceptance of diversity for the sake of adhering to multiculturalist impressions of difference (Kääpä 2014). Accordingly, the productions may be funded by specialist programs, such as the one initiated by Finnish Ministry of Education and Culture in 2012 (Suomen Elokuvasäätiö. 2012). Elina Kivihalme, the commissioner in charge of the establishment of the Finnish Film Foundation's Sámi fund, suggests this improves the Sámi industry by developing infrastructure for local production. But she also fully acknowledges that such specialist funding amounts to another case of marginalization through othering. Producing content that captures the uniqueness of Sámi cyclicality is thus extremely complex.

Considering these complexities for Sámi film production in Finland, how are the Sámi producers able to disseminate films emphasizing the cyclical perspective endemic to the culture, or ones that focus on banal everyday life, at least from the perspective of the conventional narrative logic of the mainstream film culture? The answer, at least to me, would seem to emerge from a pluriversal perspective that recognizes multiple and hybrid strategies. The emergence of a new generation of urbanized film producers addresses entirely different audiences for whom traditional notions of Sámi culture may be alien. These result arguably in the "modernist" realism of banal everydayness, but they also open the door to be considered as vital examples of a pluriversal Sámi film culture. The necessity of finding a distribution platform for such productions is helped by online networks where Indigenous film communities form as a sort of cybertarian manifestation of cosmopolitics, even as they evoke diversity under the rubric of Indigenous film culture. Yet, the picture is more complex from the perspective of those who have to work within the confines of larger film cultural constellations: "the day there is a Sámi film institute which funds Sámi language productions is the day we have the possibility of a sustainable film industry" (Simma 2015). Such a day may seem far off, but significant transformations in both thematic and infrastructural areas of Sámi film culture suggests that acceptance of the pluriversal role of Sámi culture is now more pressing than ever.

Notes

1. The Finnish Ministry of the Environment collaborated with the Sámi Council (*Saamelaiskäräjät*) in 2006 to establish a program of sustainable development for the Sámi. While much of the report indicates an inherent awareness of the potential richness of culturally diverse approaches to sustainability, the cumulative sense is that Sámi "concepts of sustainability," which privilege cyclic time and natural agency—implicit in, for example, the practice of reindeer herding and communal festivities—emerge as obstacles to achieving the Finnish government's understanding of growth.
2. For more on the politics of ecological belonging and citizenship, see, for example, Adamson and Ruffin 2013.
3. See also, Monani (2014) who discusses 'urban Indian' Jack Kohrer's *River of Renewal* (2009), where Kohrer consciously rejects the stereotype of the ecological Indian but does so in a manner that leaves much of the structural power relations of colonialism in place, key amongst them his abidance with hegemonic standards of Indigenous worthiness.

References

Adamson, Joni. 2012. "Indigenous Literatures, Multinaturalism and *Avatar*: The Emergence of Indigenous Cosmopolitics." *American Literary History* 24 (1): 143–62.

Adamson, Joni, and Kimberly N. Ruffin, eds. 2013. *American Studies, Ecocriticism and Citizenship: Thinking and Acting in the Local and Global Commons*. New York: Routledge.

Bang, Henrik, and Eva Sorenson. 1998. "The Everyday Maker: A Challenge to Democratic Governance." *Administrative Theory and Practice*, 21 (3): 325–41.

Billig, Michael. 1995. *Banal Nationalism*. London: Sage.

Chouliaraki, Lilie. 2010. "Self-mediation: New Media and Citizenship." *Critical Discourse Studies*, 7 (4): 227–32.

Columpar, Corinn. 2010. *Unsettling Sights: The Fourth World on Film*. Carbondale: University of Southern Carolina Press.

Dahlqvist, Marina. 2014. "The Attractions of the North: Early Film Expeditions to the Exotic Snowscape." In *Films on Ice: Cinemas of the Arctic*, ed. Scott

MacKenzie and Anne Stenport, 279–85. Edinburgh, UK: University of Edinburgh Press.

de la Cadena, Marisol. 2010. "Indigenous Cosmopolitics in the Andes: Conceptual Reflections beyond Politics." *Cultural Anthropology*, 25 (2): 334–70.

Finnish Ministry of the Environment. 2011. *Biologistmonimuotoisuutta koskevan yleissopimuksen alkuperäiskansojen perinnetietoa käsittelevän artikla8j: kansallisen asiantuntijatyöryhmän loppuraportti.* Helsinki: Finland. [The Final Report of a Domestic Expert Panel on a General Agreement on Biological Diversity and Article 8 Dealing with Hereditary Knowledge of Indigenous Peoples.] Helsinski, Finland.

Gauriloff, Katja, 2012. Personal Interview. October Films, Helsinki, Finland, February 15, 2012.

Ginsburg, Faye. 2011. "Native Intelligence: A Short History of Debates on Indigenous Media." In *Made to be Seen: A History of Visual Anthropology*, edited by Ruby Jay and M. Banks, 234–55. Chicago, IL: University of Chicago Press.

Harkin, Michael, and David Lewis (eds.). 2007. *Native Americans and the Environment: Perspectives on the Ecological Indian.* Lincoln: University of Nebraska Press.

Helander, Elina, and Kaarina Kailo, eds. 1998. *No Beginning, No End. The Sámi Speak Up*, Circumpolar Research Series No. 5, Canadian Circumpolar Institute.

Iversen, Gunnar. 1998. "Norway." In *Nordic National Cinemas*, ed. Tytti Soila, Astrid Soderbergh Widding, and, Gunnar Iverson, 102–41. New York: Routledge.

Kääpä, Pietari. 2014. "Northern Exposures and Marginal Critiques: The Politics of Sovereignty in Sámi Cinema." *Films on Ice: Cinemas of the Arctic*, edited by Scott MacKenzie and Anne Stenport, 45–58, Edinburgh, UK: University of Edinburgh Press.

Kailo, Kaarina. 2001. "Gender and Ethnic Overlap/p in the Finnish Kalevala." In *Of Property and Propriety: The Role of Gender and Class in Imperialism and Nationalism*, edited by Himani Bannerji, Shahrzad Mojab, and Judith Whitehead, 182–222. Toronto, Ontario: University of Toronto Press,.

Knox, Alex. 2001. "Clocks and the Sámi." University of Texas course material. Accessed May 12, 2015. http://www.utexas.edu/Sámi/dieda/natsci/clocks.htm.

Krech, Shephard. 1998. *The Ecological Indian: Myth and History*, New York: W. W. Norton.

MacKenzie, Scott and Anna Stenport, eds. 2014. *Films on Ice: Cinemas of the Arctic*, Edinburgh, UK: University of Edinburgh Press.

Mecsei, Monica Kim. 2014. "Cultural Stereotypes and Negotiations in Sámi Cinema." In *Films on Ice: Cinemas of the Arctic*, edited by Scott MacKenzie and Anne Stenport, 72–83. Edinburgh, UK: University of Edinburgh Press.

Finnish Ministry of the Environment. 2015. *Post-2015-kehitysagenda-mistä on kysymys* Helsinki, Finland.

Monani, Salma. 2013. "Indigenous Film Festival as Eco-testimonial Encounter." *NECSUS: The European Journal of Media Studies*, 1 (3). Accessed June 27, 2015. http://www.necsus-ejms.org/Indigenous-film-festival-as-eco-testimonial-encounter-the-2011-native-film-video-festival/.

———. Salma. 2014. "Evoking Sympathy and Empathy: The Ecological Indian and Indigenous Eco-activism." In *Moving Environments: Affect, Emotion, Ecology and Film*, ed. Alexa Weik von Mossner, 225–48. Waterloo, Ontario: Wilfrid Laurier University Press.

Nestingen, Andrew (2008) *Crime and Fantasy in Scandinavia: Fiction, Film, and Social Change.* Seattle: University of Washington Press.

Nixon, Rob. 2001. *Slow Violence: Environmentalism of the Poor*, Cambridge, MA: Harvard University Press.

Pentikäinen, Juha. 1989. *The Kalevala Mythology.* Bloomington: Indiana University Press.

Raheja, Michelle. 2007. "Reading Nanook's Smile: Visual Sovereignty, Indigenous Revisions of Ethnography, and Atanarjuat." *American Quarterly*, 59 (4): 1159–1185.

Simma, Paul Anders. 2015. Skype Interview. February 8.

Soila, Tytti, Astrid Soderbergh Widding, and Gunnar Iverson, eds. 1998. *Nordic National Cinemas.* New York: Routledge.

Spivak, Gayatri. 1988. "Can the Subaltern Speak?" In *Marxism and the Interpretation of Culture*, edited by Cary Nelson and Lawrence Grossberg, 271–313. Chicago, IL: Chicago University Press.

Stengers, Isabelle. 2005. "The Cosmopolitical Proposal." In *Making Things Public: Atmospheres of Democracy.* ed. B. Latour and P. Weibel 994–1003, Cambridge, MA: MIT Press.

Suomen Elokuvasäätiö [Finnish Film Foundation]. 2012. *SES:n tiedote* [The Finnish Film Foundation NewsBulletin], 19 March. Helsinki, Finland.

Suorsa, Katri, and Päivi Iskanius. 2010. *Ympäristö-, Energia- Ja Luonnonvarojen Tutkimustoiminta Pohjois-Suomessa*, (Environmental, Energy and Natural Resource Research Activity in Northern Finland). Oulu, Finland: Northern Research Information Platform (University of Oulu). http://jultika.oulu.fi/files/isbn9789514262425.pdf.

Taylor, Charles. 1997. "The Politics of Recognition." In *New Contexts of Canadian Criticism*, edited by Ajay Heble, Donna Palmateer Pennee, and J. R. (Tim) Struthers, 98–131. Toronto, Ontario, Canada: Westview Press.

Turner, Graeme. 2010. *Ordinary People and the Media.* London: Sage.

Films

Cry into the Wind. 2007. Directed by Katja Gauriloff, Helsinki, Finland: Oktober Filmi.

Curse of the Witch, The [Noidan kirot]. 1927. Directed by Teuvo Puro, Helsink, Finlandi: Suomi Filmi.

Earth is a Sinful Song, The. 1973. Directed by Rauni Mollberg, Helsinki, Finland: RM-tuotanto.

Give Us Back Our Skeletons! *[Antakaa meille luurankomme!].* 1999. Directed by Paul Simma. Inari, Finland: Sámi Film.

Kautokeino Rebellion, The [Kautokeino opprøret]. 2008. Directed by Nils Gaup. Oslo, Norway: Rubicon.

*Last Yoik in Sámiland [Viimeinen joika Saamenmaalla].*2008. Directed by Hannu Hyvönen, Iisalmi, Finland: Elonmerkki.

*Minister of Sagojoga, The [*Ministern från Sagojoga.]. 1997. Directed by Paul Simma. Stockholm, Sweden: Moviemakers Sweden.

Pathfinder [Ofelaš]. 1987. Directed by Nils Gaup: Oslo, Norway: Filmkameratene.

River of Renewal. 2009. Directed by Carlos Bolado. Berkeley, CA: Pikiawish.

Tale of Arctic Love, The [Iskko-Máhte máinnas]. 2000. Directed by Paul Simma, Uppsala, Sweden: Firstfilm and Television.
Umur. 2002. Directed by Kai Lehtinen. Helsinki, Finnland: Fantasiafilmi.
White Reindeer, The [Valkoinen Peura]. 1952. Directed by Erik Blomberg, Helsinki, Finland: Junior Filmi.
Wolf [Varg]. 2008. Directed by Daniel Alfredson. Helsinki, Finland: Lumi Filmi.

8 Resistance and Hope in Mohawk Cinema

Iroquois Cosmologies and Histories

Shelley Niro and Salma Monani

Introduction: When the Personal is Cosmopolitical (by Salma Monani)

Shelley Niro's cinematic work, an audiovisual conduit into which she melds her multimedia interests in all forms of creative expression—from dance and music to painting, photography, and artisan craftsmanship—spans over twenty years. It is celebrated and has received numerous prestigious art awards, not only for its innovative expression but for its statements on the complex interweave of the personal and political in contemporary Indigenous experiences.[1] As Lawrence Abbott writes in his well-cited "Interviews with Loretta Todd, Shelley Niro and Patricia Deadman" (1998), "the interplay of the personal and the political, the mixing of the minutiae of daily life with the broader forces that impact that life is one of the underlying structures of Niro's work" (349). Niro's work is strongly inspired by a need to personally talk back and through political marginalization and trauma—historical and contemporary.

At the same time, what makes Niro's work particularly valuable as an *eco*cinematic archive are its *cosmo*political dimensions. Niro draws from her Iroquois communities' cosmologies, which as she suggests from her following reflections blend nature with culture. Such cosmovisions are clear in her most explicit environmental film, her mid-career creation, *Tree* (2006).

A five-minute, black-and-white, artistic piece shot on 35mm film, *Tree* was commissioned by LIFT, the Liaison of Independent Filmmakers of Toronto, and though Niro was given complete artistic freedom in her project, she chose to represent, "a day in the life of Mother Earth." Putting her own twist on the iconic "Crying Indian" campaign advertisement ("Crying Indian" 1978), Niro presents her protagonist as Mother Earth, who visits a contemporary city and is saddened by what she sees—a cold, unpeopled, and lonely place.

Unlike the ad, which uses a non-Native actor (Iron-Eyes Cody) for what many have critiqued as a disingenuous campaign to forward a consumerist, environmental agenda with little regard for Indigenous concerns beyond a stereotypical trope (e.g., Dunaway 2015; Raheja 2010), Niro's Mother Earth, played by Anishnaabe artist Lena Recollet, centers Indigenous perspectives in distinctly Iroquois ways. Specifically, Niro represents Mother Earth as a spirit-being who is simultaneously human and nonhuman and thus challenges binary divisions of nature and culture.

Figure 8.1 Still from *The Tree* (2006). Lena Recollet (Anishnaabe) plays the hybrid human-tree spirit of Mother Earth. Source: Courtesy of Shelley Niro.

While binaries have often been the way Western cultures frame the world, with nature as inarticulate object without political say on one side, and culture as the realm of people (with some having more right to political power than others), traditional Iroquois cosmologies resist such binaries.[2] Here, Mother Earth is daughter to the creator of the Iroquois world, Sky-woman, and mother to the twins—day and night—from whom the matrilineal clans of the Iroquois people descend. As Haudenosaunee Elder Tom Porter (2006) describes, the Iroquois people emerge from earth and sky and their traditional rituals return the dead back to the earth and sky. Throughout their lives, traditional Iroquois rituals, like tobacco burning, which enliven the senses, remind the living of their inseparable links to these material and ethereal worlds.[3]

As Marisol de la Cadena (2010), Adamson (2012), and others in this volume write, Indigenous peoples' recognitions of natureculture continuums are hard-fought, as Western binaries and hegemonies often dismiss such traditions as mere superstition. Niro's *Tree* captures a historical sense of such dismissal in the way Mother Earth, a representative of Iroquois ways of being, is confronted by a degraded, deserted cityscape. Instead of the lively, peopled, more-than-human ancestral lands with which she (and we as viewers) might interact, she encounters a lonely, shattered world. By using not only local shots of Toronto but also globally recognizable shots (for example, the memorial dome at Hiroshima), *Tree* extends its political commentary on the destruction wrought on Iroquois life-ways to a planetary scope. The film, which ends with a distraught Mother Earth morphing back into a tree, can be read as multilayered.

On the one hand, it speaks to how Western hegemonies have relegated Indigenous cosmovisions passive and inarticulate in a callous, cold world. On the other hand, it draws attention to the resistance and response of Iroquois and nonhuman agents in a world dominated by Western-centric epistemologies of reductionism. As Niro describes in an interview with Mike Hoolboom (Niro 2008, 100), "It's about nature's response, which is occurring now with global warming, for instance. At the end she [Mother Earth] has this great look in her eyes. I love that part. Where her eyes are so intense. It's about humanity and the environment."

In drawing together humanity and the environment, Niro's Iroquois sensibilities turn her cinema into a space for cosmopolitics, where the screen is "is peopled with shadows of that which does not have, cannot, or does not want a political voice" (Stengers, 995). As the three examples she reflects on below suggest, her work is deeply committed to retrieving the voices of those who have been relegated to the shadows by 500 years of colonial history and the continued violence of neocolonialism. Each film is inspired by a political event that is indicative of eco-social trauma and crisis. In responding to these events, Niro centers the Iroquois and their entangled relations to the more-than-human world in ways that give us a sense of how her cosmopolitics are richly complex. Using stills from *It Starts with a Whisper* (1992), *Honey Mocassin* (1998), and *The Shirt* (2003), Niro illustrates how her work is grounded in: (a) experimentation with multimedia use, (b) play with tone (from seriousness to humor), (c) feminist and queer themes, and (d) insistence on acknowledging community. Prefaced by reflections on her childhood, her commentary makes us aware that Niro's personal experiences have always been cosmopolitical. Each of her artistic choices prompt her viewers, both Indigenous and non-Indigenous, to encounter a world that questions boundaries—of past/present/future, material/ethereal, self/other, and as a consequence of the very ideas of nature/culture.

Childhood Influences (by Shelley Niro)

I was born in Niagara Falls, New York, in 1954. I lived there until 1959. I was living an urban Indian existence; movies and television played an important part in my early childhood. Living in the outdoors was not a big part of my early life. *Wagon Train*, *Stagecoach West*, *I Love Lucy*, *The Three Stooges* and much, much more were some of the daily shows I saw. Real mind expanders for a four-year-old. My mother liked the convenience. I liked the constant flow of images, dialogue, and soundtrack. It was exciting.

During those early years I was not privy to what propaganda was and how it would form whole nations of people's opinions about the land they were living on and how they got it.

When I was five, my family moved back to the Six Nations Indian Reserve, Ontario, Canada. The community we came to know was kind and thoughtful. Acceptance of others was the norm. There was never any question of intolerance. My memory serves that the elders were nonjudgmental and this was also reflected in members of the community around me.

We lived on the fourth line of the Six Nations Reserve. Life here was frugal. We had to get water from a well. We had a wood-burning stove and used kerosene lamps.

The road was dirt, and the trees lined both sides that stretched on forever. Living here had its benefits. We could play from morning to night in vast fields and forests. Children who lived in the area played as a group. Nature was bountiful. Frogs, birds, bugs, and snakes were part of our daily observations. The winter was a time of solace and quiet. I believe this solace brought with it a desire to communicate with the outside world, to know what was happening, and to be entertained.

Figure 8.2 House On Fourth Line. Source: Courtesy of Shelley Niro.

When I was six, all of the trees were cut down to make way for hydroelectric poles that soon would dominate the never-ending road. The road itself was also going to be broadened. Eventually it was paved. It was a dramatic change from large Carolinian trees to nothing. This change also meant we could hook up to the hydro-lines and become a member of the viewing masses once again.

At the time, the majority of entertainment took the form of films, namely Hollywood Westerns. They were everywhere and seemed to be on all of the time. I wasn't too impartial to these kinds of movies. I had to get my adrenalin rush somewhere. But, as time went on I became sensitive to this ongoing portrayal and the never-ending slaughter of Indians, yelling, screaming, and whooping, on horseback.

Though we weren't from the western lands portrayed in Hollywood, much of mainstream culture saw us all as one type of "Indian." Historically, the Iroquois community has been seen as killers. A bunch of psychopaths lying in wait for an opportunity to strike again.

However, within the last twenty-five years, especially since the cultural revival associated with the civil rights movement of the 1960s and the fervor of the 1970s, as well as the advent of lighter, easier-to-use video cameras, Native filmmakers from all over America have resisted this harmful and homogenous stereotype. From the silly to the sublime, we see ourselves and feel power within the image projected onto a screen in the traditional setting of a theater, a community hall, or any place that has an electrical power outlet. We draw on our cosmologies and traditions to ground our work and celebrate intellectuals, teachers, film producers, and our own historians in telling our stories. In all my own work, I seek to highlight the wellspring of intellectual and cultural creativity in Iroquois pasts and presents as a means forward.

Cinema as a Project of Hope (by Shelley Niro)

It Starts with a Whisper (27:30 Minutes Duration, 1993)

The year 1992 is significant, for it was the year the world was celebrating the quincentennial anniversary of Christopher Columbus' discovery of the Americas. Museums, galleries, and institutions of education were commissioning exhibits to represent this moment. Artists, writers, and historians were creating works (re)telling Columbus's journey and all that it brought to the world's stage, such as wealth, resources, slavery, and land claims that mostly benefited those of European descent.

As a descendent of America's First Nations, for this moment I wanted to make something that remembered the event and its ensuing legacy in a different way. I also wanted to make something besides a painting that hung on the wall momentarily, or a sculpture that would only be temporarily prominent.

It Starts with a Whisper has my first major cinematic effort. I worked with Anna Gronau, who was a practiced filmmaker. We filmed at six different locations on the Grand River, which flows through the Six Nations Reserve, and engaged an all-Native cast. It is a short film, set in contemporary times, and features a young seventeen-year-old Mohawk girl, Shanna Sabbath.[4] She is having trouble adjusting to her young life. She is burdened by the history of Native American people. Her mind is filled with facts and statistics of Native communities plagued by suicide, poverty, unemployment, mental health issues, and general unhappiness connected to the demise of many of the Native nations in North America.

With her three aunts, she is headed to celebrate New Year's Eve at Niagara Falls, which is ancestral land and spiritually important to the Iroquois people. Shanna's aunts are in a partying mood. As soon as Shanna gets into the car, they tease her lovingly and try to change her mood. She resists their camaraderie. As soon as they get to Niagara Falls, she runs away from them, and finds herself stepping into a dream world.

Figure 8.3 Still from *It Starts with a Whisper* of Debra, Beverly, and Elizabeth Doxtater, the filmmaker's sisters, playing Shanna Sabbath's fun loving aunts. Source: Courtesy of Shelley Niro.

Here, the names of the many Native nations roll through her head. She is overcome, but then respected Oji-Cree elder Elijah Harper appears as an ancestral spirit. As Canada's first Treaty Indian to be elected as a provincial politician, Harper had single-handedly stopped the 1990 Meech Accord, an attempt at constitutional reform, by refusing to give his vote for the consensus required by Canada's parliament.[5] His refusal was strongly supported by Indigenous communities who agreed that the Accord's negotiations had denied Canada's First Nations input. In *It Starts with a Whisper*, Elijah gives Shanna advice, helping ease her worries and giving her hope.

Figures 8.4 and *8.5* Stills from *It Starts with a Whisper.* Tobey Burning played Shanna Sabbath, while Elijah Harper (Oji-Cree) plays an ancestral spirit. Source: Courtesy of Shelley Niro.

To welcome 1992, the film was first shown New Year's Eve, in The Lodge, a senior center on the Six Nations Reserve. Its premiere included over 200 community members in the audience, and celebrated, that yes, we have survived the last 500 years. The film's last scenes alert us that it is 11:55 P.M., December 31, 1991, as Shanna returns to her present time and gathers with her aunts.[6] In their celebration, one aunt pours tea from a pot into cups, one cuts a cake, and one recites an E. Pauline Johnson poem. Shanna hits a waterdrum and sings a song. Fireworks cover the sky, turning into beadwork of the Iroquoian Celestian Tree.

Figure 8.6 Still from *It Starts with a Whisper.* Iroquois cosmology, represented by beadwork, graces the New Year's night sky at the end of the film and the start of 1992, the 150th anniversary of Christopher Columbus' "discovery" of America. Source: Courtesy of Shelley Niro.

Honey Moccasin (43 Minutes Duration, 1998)

Set in contemporary times, this is a story about communities who are trying to recover their collective histories through stories and music.[7] The main protagonist, Honey Moccasin (Tantoo Cardinal) owns a bar/cafe, which serves as a community center, gathering place, and town hall of sorts for the residents of the Grand Pine Reserve. They sing, eat, discuss the topics of the day and entertain each other, including by using the stage where performance art is shown, but not always understood.

Figure 8.7 Still from *Honey Mocassin.* Tantoo Cardinal as Honey.
Source: Courtesy of Shelley Niro.

For example, Mabel (Florene Belmore), Honey's daughter, uses her mother's bar customers for her audience. In one of her performances, Mabel makes a teepee with a hole that she sticks her face through.[8] She sings Peggy Lee's song Fever while black and white photos screen on the outside of the tent. The images are of Native people photographed in the late part of the 19th century and early part of the 20th century. These are mixed in with photos of early Canadian soldiers. Other photos show residential school children with nuns. The images evoke a sense of colonial authority and martial presence as part of the landscape.

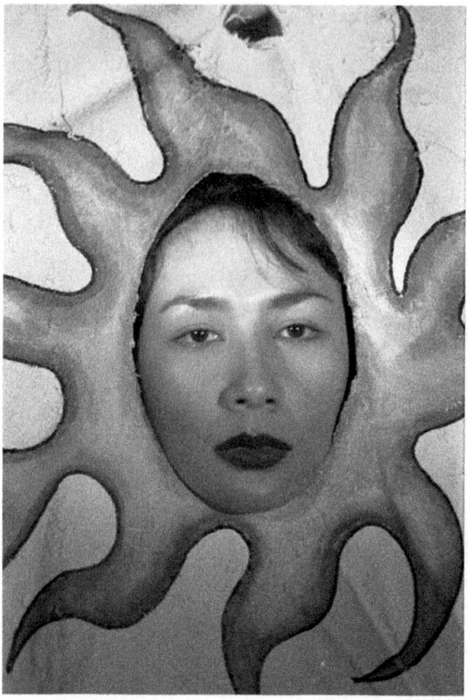

Figure 8.8 Still from *Honey Mocassin*. Mabel (Florene Belmore) during one of her mixed media performances at her mother's bar. Source: Courtesy of Shelley Niro.

One of the lines to the song is "when you put your arms around me, you give me a fever that will surely burn." There are quick shots of a person with small pox. In the film, the audience watches the performance, not really sure how to react. With great encouragement from Honey, they appreciate Mabel's attempt. Mabel's work takes on added significance, as she crafts another film that helps tie-up *Honey Mocassin's* plot of the mysterious theft of powwow outfits.

With the upcoming powwow, everyone has been in a panic when they can't locate their outfits. One of the elders, Johnnie John (Tom Hill) tries to settle the reserve people by saying, "yes, these things have been taken but we can make new ones; using our brains and the materials we have at hand, we can make better ones." This moment is in direct reference to Native American resistance to colonial theft of local resources through broken treaties, and harmful policies of extermination and assimilation that destroyed communities.

While the residents heed Johnnie John's advice and begin to make new outfits with gusto and passion, Honey sleuths to find the thief, who she uncovers as Two-spirited Zachary John (Billy Merasty), son of Johnnie John. He's been hiding the outfits into his basement, where he tries them on, teasingly playing the part of femme fatale, but also questioning his worth as he looks into the mirror.[9]

Figure 8.9 Still from *Honey Mocassin*. Zachary John (Billy Merasty) in powwow drag. Source: Courtesy of Shelley Niro.

When he is discovered by Honey, Zach attempts to kill her, but is stopped and soon regrets his foolishness, recognizing that his community is willing to forgive and accept him for who he is. Specifically, Mabel casts him in her film *Inukshusk*, which is a performance piece based on a poem by First Nations poet and playwright, Daniel David Moses.[10] After the screening of *Inukshusk* in Honey's bar, Zach is applauded by all. He learns to accept himself, and his father hugs and accepts him too, forgiving him for stealing the outfits.

Honey Moccasin is a creation of characters who had never been seen before in mainstream cinema. It was also meant to be humorous and mindful of First Nations history. I developed its many layers to articulate the complexity of Native experiences.

The Shirt (5.55 Minutes Duration, 2003)

The Shirt, a satirical short, is a five-minute, no-budget video. I conceived it as I flew over Texas and saw the land below, parceled, privatized, and maintained. That was in 2002. In 2003, the President of the United States, Texan George W. Bush, was sending troops to Iraq. The first thought that came to mind was that Iraq's sheep herders, and others who lived off the land, would not have a clue about what would be happening as bombs

Figure 8.10 Still from *The Shirt*. Hulleah Tsinhnahjinnie plays the woman without a name, with a shirt, and then, with no shirt. Source: Courtesy of Shelley Niro.

fell. Military personnel would soon be descending on the vast areas of land around them, causing mayhem and general destruction in their lives and in those of their families for generations to come. Much like what had happened to Native Americans, changing their lives forever.

The Shirt is filmed in the suburbs of Irvine, California, and along the Grand River, in and around Niagara Falls. I asked my good friend Hulleah Tsinhnahjinnie and her life partner Veronica Passalacqua if they would participate. They said they would. ElizaBeth Hill composed the soundtrack, which is gentle in its rhythmic chants and beat. It intersplices tracking and panning images of these landscapes and waterscapes with that of Tsinhnahiinie standing in the foreground of green hills. She wears a U.S. flag bandana on her head,and a white t-shirt, whose letterings spell out the story of Native Americans dispossession. The spellings end with the shirt revealing, "and all I get is this shirt." Then we see Tsinhaniinie standing without her shirt. At the very end, Passalacqua is also shown standing with the "and all I get is this shirt" t-shirt.

Conclusion (by Salma Monani)

Niro's cinematic archive is rich and broad. Yet, as suggested by the examples above, each of her uniquely distinct artistic endeavors is grounded in common themes—a political spark, a sense of humor with which to look past pain towards a more hopeful future, and an insistence to recognize the communities that assist in this process of resistance, recovery, and rejuvenation. Because community is defined by Niro's Iroquois sensibilities, it both emphasizes Iroquois agency and draws attention to how Iroquois life-ways blur natureculture boundaries, whether through *Tree's* spirit-being Mother Earth, the presence of sacred land (Niagara Falls) as necessary material grounding for Shanna's link with her ancestors in *It Starts with a Whisper*, *Honey Mocassin's* heteronormative gender critique, or land as ultimate resource in *The Shirt*. When placed in the context of Niro's larger corpus of work, which is often inspired by values inscribed in the story of Hiawatha and the Peacemaker, this sense of enlarged community takes on additional cosmopolitical meaning.

As Niro explains of the story, which is a central event in Iroquois cosmology and history, "the legend begins in a deep period of grief." Hiawatha and the Peacemaker had to counter terrible conflict and social upheaval to create a democratic transnational alliance that helped unite the original five nations of the Iroquois Confederacy to provide a particularly resilient front against the growing onslaught of European influence. As I have written in my analysis of her 2009 feature, *Kissed by Lightning*, Niro articulates the value

Figure 8.11 Still from *Kissed by Lightning*. Mavis Dogblood (Kateri Walker), Bug (Eric Schweig), and the filmmaker's paintings of Haiwatha and the Peacemaker. Source: Courtesy of Shelley Niro.

of this story of the "Great Law of Peace" as quintessentially cosmopolitical (Monani, 2014). It speaks to the possibilities of living through hardship but it does so with attention to relations "based on care and responsibility for land that all can share" (Smith, quoted in Monani, 2014, 140). As Niro explains, the Peacemaker helped his people eat the right foods, versus being cannibals, helped them use tobacco for peace, and he taught them ceremonial singing. Through these actions, he helped ground his people in physical senses that enmesh material and mental processes of thoughtful engagement with human and more-than-human alike.

Niro's films, which draw from the Peacemaker story and other Iroquois sensibilities, enact a similar process of cosmopolitical grounding. From those highlighted here, to others like *Suite: INDIAN* (2005), *Rechargin'* (2007), and *Robert's Paintings* (2011), with their attention to the material affect of music, dance, and painting as regenerative motifs of Iroquois ways, Niro cinematic expressions serve as interventions of hope, resistance, and resilience in ongoing histories of eco-cultural genocide and trauma.

Notes

1. Recognizing her exceptional artistic talents, the Ontario Arts Council named her the first laureate for their Aboriginal Arts Award (2012). Niro has received many awards over the years, including being made a fellow of the Smithsonian's National Museum for the American Indian, Washington, DC (1997).
2. The "Crying Indian" campaign, as part and parcel of Western environmental politics, suggests this distinct binary too—nature is an object used by people. Unlike Niro's personified Mother Earth, it has no subjective agency.
3. As I discuss at the close of this chapter, Niro uses the Peacemaker story to show that materiality is intrinsic to evoking thoughtful engagement with the world.
4. Tobey Burning played the main character, Shanna; she was also the set and wardrobe designer. My three sisters, Debra Doxtater, Beverly Doxtater and Elizabeth Doxtater, act in the film. The music for the film was also composed by family and friends (my brother, Michael Doxtater, as friends, James Miller and Sadie Buck).
5. More about Elijah Harper and his prominent role can be found on the Canadian Encyclopedia entry (Marshall 2015). His presence is still strong as is evident from a recent CBC story that discusses his 25-year legacy (2015).
6. While the film ends right before midnight, we had to be out by 8pm as the elderly folk had to go to bed by then. We were fine with this and screened the film earlier.
7. For example, the film features Indigenous singer-songwriter and acoustic guitar player Willie Dunn, as the character Hank. I humbly asked Mr. Dunn if he could compose music for *Honey Moccasin*. Much to my indebtedness, he accepted and wrote beautiful songs for the film. At the end of the film, he is shown getting escorted out of a moving van, continuing to play his guitar and sing his song. While Dunn passed away in 2013, his important work can still be found on sites such as the Canadian National Film Board (see for example, http://www.nfb.ca/film/ballad_of_crowfoot).

8. The artwork for *Honey Moccasin* was created by a class of students I had while facilitating a video production program at the University of Western Ontario.
9. Most of the outfits for Zachary John's basement scene were borrowed from a local Six Nations powwow dancer, Margaret Key. They were authentic beaded outfits.
10. Daniel David Moses is a well-recognized playwright and poet from Canada's First Nations (Delaware descent). As an openly gay writer, Moses speaks eloquently to issues of gender and sexuality; his work is also richly threaded with themes that speak to the interconnections between the natural world and humanity, as is his poem "Inukshuk" (1990). You can find more information about Daniel David Moses' work at his website: http://www.danieldavidmoses.com/.

References

Adamson, Joni. 2012. "Indigenous Literatures, Multinaturalism, and *Avatar*: The Emergence of Indigenous Cosmopolitics." *American Literary History* 24 (1): 143–62.

CBCNewsStaff.2015."25YearsSinceElijahHarpersaid"No"totheMeechLakeAccord. *Canadian Broadcasting Company News*. June 11. Accessed January 2, 2016. http://www.cbc.ca/news/canada/manitoba/25-years-since-elijah-harper-said-no-to-the-meech-lake-accord-1.3110439.

"Crying Indian (The): 'Keep America Beautiful' Commercial." 1978. *NBC Nightly News*. New York: NBC Universal, NBC Learn Higher Ed. 15 October. Accessed Fri Jan 23 2015. https://highered.nbclearn.com/portal/site/HigherEd/browse/?cuecard=36227.

de la Cadena, Marisol. 2010. "Indigenous Cosmopolitics in the Andes: Conceptual Reflections beyond 'Politics'." *Cultural Anthropology* 25 (2): 334–70.

Dunaway, Finis. 2015. *Seeing Green: The Use and Abuse of American Environmental Images*. Chicago: University of Chicago Press.

Marshall, Tabitha. 2016. *Canadian Encyclopedia: Historica Canada*. s.v. "Elijah Harper." Accessed January 2, 2016. http://www.thecanadianencyclopedia.ca/en/article/elijah-harper/.

Monani, Salma. 2014a. "*Kissed by Lightning* and Indigenous Cinema's Natureculture Continuum." In *Ecoambiguity, Community and Development*, ed. Swarnalatha Rangarajan, Vidya Sarveswaran, and Scott Slovic, 131–47. New York: Lexington Press.

Moses, Daniel David. 1990. "Inukshuk." In "Native Writers and Canadian Writing." Special Issue, *Canadian Literature* 124–25: 241–42.

———.n.d. Personal website. Accessed January 2, 2016. http://www.danieldavidmoses.com/.

Niro, Shelley. 2008. "The Red Army Is the Strongest." Interview with Mike Hoolbloom. *Practical Dreamers: Conversations with Movie Artists*. ed. Mike Hoolboom. Toronto, Ontario: Coach House Books.

Porter, Tom. 2006. "Mohawk (Haudenosaunee) Teaching." Canada: Four Directions Teachings. Accessed December 23, 2015. http://www.fourdirectionsteachings.com/transcripts/mohawk.pdf.

Raheja, Michelle. 2010. *Reservation Reelism: Redfacing, Visual Sovereignity, and Representations of Native Americans in Film*. Lincoln: University of Nebraska Press.

Stengers, Isabelle. 2005. "The Cosmopolitical Proposal." In *Making Things Public: Atmospheres of Democracy*, ed. Bruno Latour and Peter Weibel, 994–1004. Cambridge, MA: MIT Press.

Todd, Loretta, Shelley Niro, and Patricia Deadman. Abbot, Lawrence. "Interviews with Loretta Todd, Shelley Niro, and Patricia Deadman." *The Canadian Journal of Native Studies* 16 (2): 335–73.

Films

Ballad of Crowfoot, The. 1968. Directed by Willie Dunn. Songs written by Willie Dunn. Canadian National Film Board, 2016. http://www.nfb.ca/film/ballad_of_crowfoot.

Honey Moccasin. 1998. Directed by Shelley Niro. Performances by Lynn Hutchison and Tantoo Cardinal. New York: Women Make Movies.

It Starts with a Whisper. 1993 Directed by Shelley Niro and Anna Gronau. New York: Women Make Movies.

Kissed by Lightning. 2009. Directed by Shelley Niro. Performance by Kateri Walker. October. Toronto, Ontario: Turtle Night Productions.

Rechargin'. 2007. Directed by Shelley Niro. Distributed by VTape.

Robert's Paintings. 2011. Directed by Shelley Niro. Distributor VTape.

Shirt, The. 2003. Directed by Shelley Niro. Performance by Hulleah Tsinhnahjinnie. Toronto, Ontario: VTape.

Suite: INDIAN. 2005. Directed by Shelley Niro. Distributor VTape.

Tree. 2006. Directed by Shelley Niro. Performance by Lea Recollet. Toronto, Ontario: LIFT (Liaison of Independent Filmmakers of Toronto).

Part III

Multispecies Relations

9 A Network of Networks

Multispecies Stories and Cosmopolitical Activism in *Solar Storms* and *People of a Feather*

Kathryn Yalan Chang

Biologist and filmmaker Joel Heath spent seven winters filming the eco-documentary, *People of a Feather* (2011), which focuses on how the eider duck, a species now facing extinction, and the Inuit living on the Belcher Islands in Canada's Hudson Bay have suffered from climate change due to hydroelectric energy development. By using underwater cameras, time-lapse photography, and satellite imagery, *People of a Feather* reveals how massive dams that power New York City and other eastern North American cities affect marine environments and wildlife. What the film shows is how hydro-electric dams are changing ocean currents and having an adverse effect on sea ice ecosystems in Hudson bay. This devastatingly informative film, when used to contextualize Linda Hogan's novel, *Solar Storms* (1995), shows that today, twenty years after publication of the novel, fiction and film can work together to alert the public to risks in accessible ways not necessarily common to scientific articles and technical reports.

Solar Storms is based on the impacts of a large dam and fictionalizes events surrounding actual construction of a hydroelectric power plant in James Bay, Canada. The novel traces the journey of main protagonist Angel and her grandmothers north to their ancestral homelands to find Angel's birth mother. On the way they encounter animals and plants as beings that are intimately entangled in and dependent on the ecosystems of which they are a part. In the "swamp roots" (Hogan 1995, 308), Angel sees a kind of "grassroots organizing," or what I will call in this chapter "a network of networks" among humans, waters, lands, plants, and animals.[1] To articulate this concept, in the first section of the chapter I engage in a material, ecocritical analysis that focuses on how Hogan's *Solar Storms* anticipates *People of a Feather*, which in turn documents dam-building as a force of "globalization from above, to use cultural anthropologist Arjun Appadurai's phrase (2001a, 19). The second section examines *Solar Storms*'s broad decolonizing themes. I explore how Hogan tells the "stories of a pluriverse"—where all things matter and are networked. Hogan's waterscape and cosmology are woven into "dreams of Earth" (Hogan 1995, 170) and composed of what Iovino and Oppermann have called "storied matter" (2012). Hogan shows why Native American oral storytelling cannot be separated from geographical locations (or actual physical places) and from cosmological identities. By putting emphasis on "cosmopolitics" (Adamson 2012b; de la Cadena 2012), a politics that takes all species' survival into consideration, the last section tackles how people exercise their grassroots solidarity to resist and to fight for their survival by activating the spirit of "trans-indigeniety" (Allen, 2012) supporting this growing "network of networks."

A Material Ecocritical Reading of the Dam Project

Solar Storms is recognized by literary critics and ecocritics as an important artistic representation of the connections between social and environmental issues. Most of the criticism on the novel discusses the significance of place, mother land, remembrance, origins, beginnings, the ecological Indian, narrative strategies, colonialism, and environmental justice issues.[2] This section argues Hogan's *Solar Storms* anticipates the unpredictable risk of a dam project, the actual consequences of which, twenty years after publication of the novel, are documented in *People of a Feather*.

In *Solar Storms*, Linda Hogan writes about the struggle of a seventeen-year-old mixed-blood girl, Angel Wing, and her (grand)mothers, Bush, Agnes, and Dora-Rouge. To write the stories of these women, Hogan fictionalizes the histories of the Cree and Inuit peoples who organized to fight construction of the James Bay hydroelectric project in Quebec, Canada, which began construction in 1971. *Solar Storms* shows how First Nations people had lived in place for thousands of years. However, after construction of the dam began, their long ecological relations were disrupted. Hogan reveals how an interconnected multispecies world rapidly began to degrade, and in the process, threatened the local people's subsistence lifeways.

People of a Feather documents the nonfictional impacts of the dam on First Nations, and more specifically on the Inuit people, by examining how the marine environment plays a crucial role in everyday Inuit life. Like the documentary, *Solar Storms* shows how water-and-human relationships figure materially in the lives of Angel and her relatives. On Angel's journey through the traditional lands of her people, Angel feels her body remembering landscapes and waterscapes, even though she had left these places behind when she was a baby and is now returning years later, as she turns seventeen. Both novel and documentary illustrate that the human body experiences memories through its senses (of taste, touch, smell. etc.). Such that interactions with the environment invoke experiences with "storied matter." As material ecocritic Serpil Oppermann has argued. "Elements, cells, genes, atoms, stones, water, landscapes, machines, among innumerable others, are embodied narratives, repositories of storied matter" (Oppermann 2013, 59).

The notion of "storied matter" invites us into "a polyphonic story" that includes "the vital materiality of life, experiences of nonhuman entities, and our bodily intra-actions with all forms of material agency as effective actors" (Iovino and Oppermann 2012a, 88). *Solar Storms* demonstrates that material processes like the construction of a mega-hydroelectric project can take on a form of storied agency, causing displacement of Indigenous people and leading to poverty and misery. Moreover, when understood in terms of globalization, the agency of these dams helps to reveal their environmental-political complexities. As Donald A. Grinde and Bruce E. Johansen indicate about the historical consequences of the James Bay projects, "It is important to understand that the James Bay projects are not single dams across single rivers that flood valleys between mountains. They are massive, earth-moving projects across an area as large as the state of Oregon [that impact the Cree and Inuit peoples]" (Grinde and Johansen 1995, 225). Arundhati Roy has argued, "Big dams are to a nation's 'development' what nuclear bombs are to its military arsenal. They're both weapons of mass destruction. They're both weapons governments use to control their own people" (1999, 31). *Solar Storms*, as fiction, represents the consequences of big dam development on both humans and nonhumans. Hogan shows how a whole landscape is erased and, afterwards, tribal people's subsistence life is rendered impossible. Although dams correlate to Western Enlightenment thinking about the benefits of "development" and "progress," they cause social and environmental injustices on a massive scale. For example, connecting dam construction with European colonizers, Vandana Shiva, who opposes neoliberal dam construction projects, explains, "For European colonizers who came to America, river colonization was a cultural obsession and an imperial imperative. Nature in general, and rivers in particular, were valued for their commercial benefit and were seen as being in need of taming" (2000, 53–54). Hogan represents in her novel the devastation that accompanies such dam-building in *Solar Storms*.

In the novel, Angel, who was severely scarred on her face as a baby, witnesses the destruction of her people, women and children, and landscapes (waterscapes) at the advent of the dam construction between the United States and Canada. Historically, as "the first mega-scale hydro project" (Hornig 1999, i), the development of Quebec's hydroelectric resources at James Bay was immediately "highly controversial" (Coppinger and Ryan 1999, 41). While the non-Indigenous settlers in the novel see the construction of dams as "progress" that replaces the backwardness of "savages," Angel and her family see the construction as the end of all familiar landscapes/waterscapes and of the trees, animals, and so forth that they and their ancestors have known for centuries (Hogan 1995, 284). In *Solar Storms*, the scars on Angel's face symbolically represent the way dam construction irreparably damages "the land, the water, the rivers" (268). Dams do not just change the appearance of a place, but they are "eating land, eating people, eating tomorrow" (302).

Hogan's prediction of the irreversible damage of the dam project anticipates the actual events shown in *People of a Feather*, which are happening today to the Inuit people living on Belcher Islands in Canada's Hudson Bay, to the eider ducks, and to the overall ecosystems on which they have depended for their survival and sustenance for thousands of years. As documented in the film, the size of the dams and the reservoirs behind them has had a huge impact. One "single reservoir covers almost 3,000 square kilometers of drowned forest and river valley. Phase one flooding extends 600 kilometers inland ... in a chain of huge reservoirs that would more than bury Prince Edward Island" (42:40–43:06). Indeed, the James Bay project in northern Quebec could be compared in scale with other dam developments, including "the Narmada dam in India, the Three Gorges dam on the Yangtze in China, the Challawa Gorge dam in Nigeria, and the Xingu project in Brazil" (Warner and Coppinger 1999, 19).

Such mega-scale hydro projects not only impact local people's lives and environments but also ensure that Indigenous people will not be "the principal beneficiaries of the proposed development" (Warner and Coppinger 1999, 19). As residents of the Great Lakes basin between the United States and Canada understand, the James Bay Hydroelectric Project also caused tremendous environmental impacts in the form of contamination of waters. The director of *People of a Feather*, Joel Heath, documents how the Inuit people's lives have been deeply impacted by the dam's construction. According to the Inuit people interviewed in the film, "In the 1970's, things started to change. Many rivers were dammed ... trapping our freshwater in reservoirs" (28:17–28:31); "For almost forty years now ... reservoirs have dumped [our] water onto the sea ice habitats of Hudson Bay ... at the [wrong] time of year. Our people on the Belcher Islands were some of the first to notice the consequences" (28:46–29:03). Because of the freshwater the massive dam project brings into the bay, both Inuit and eider ducks confront changing currents, which threaten their survival during harsh Arctic winters. This change to the circulation of waters cannot be seen by the human eye, however, it is documented by *People of a Feather*.

In *Solar Storms*, there is a long canoe trip taken by Angel that predicts how the circulation of water will be interrupted by a dam. Angel and her grandmothers also witness how the land is ravaged by the rising waters. Hogan bases these scenes on what happened in the first phase of the James Bay hydroelectric project. According to Grinde and Johansen, who wrote about the immediate impacts after the project was launched, "Traditional family patterns and ways of making a living have been breaking down; one-quarter of the Crees' caribou herds, about twelve thousand animals, drowned in the first phase of the project ... The entire culture and cosmology of the Crees revolves around seasonal activities associated with the land and its waters" (Grinde and Johansen 1995, 223). Even in these early stages it was clear that dam construction was destroying not just the material sustenance of Native American lives but also their spiritual lives, cultures, arts, languages, and cosmologies.

Dam construction is one example of a force that Arjun Appadurai has termed "globalization from above" (2001b, 19). This process reduces a sense of place by making all space homogeneous; one space comes to look like other places in the world that have also had dams constructed nearby. Hogan writes that houses in Adam's Rib (Angel's village) begin to appear to the outsider as ambiguous "dark brown and square, with nothing to distinguish [them] from the others" (Hogan 1995, 27). The homogenizing effect of the developer's imperial imagination legitimizes and imposes a standardized geographical vision on the landscape. In the creation of this space of Western "progress", Indigenous cosmologies and creation stories are undermined. As Marisol de la Cadena points out, the pluriverse (multiple worlds) begins to disappear; only a single world (universe) remains under Western colonization (2010, 345).

To further articulate this loss, Hogan calls attention to the toxic/poisoned landscape by the name she gives to Angel's village in the novel. At the outset of the story, Angel stays in Adam's Rib, an impoverished village where her (grand)mothers, Agnes and Dora-Rouge, reside. These women tell her how the "first women at Adam's Rib called themselves the Abandoned Ones" (Hogan 1995, 28), because they and their children had been left behind by their men, who moved on to other places for more fur trade when the land was worn out. From these names, "Adam's Rib" and "the Abandoned Ones," the implication is that White men (here represented as French fur trappers) exploit the land, and after it is used up desert the women and the place. What is left behind is desolated land, in which women have to deal with the poverty. Another example is "Poison Road." The French named it *Poisson*, "because once it had rained tiny fish onto this road" (Hogan 1995, 24). However, after the European settlers' toxic activities, the name is changed from "fish" to "poison," words that signal a transition from the original abundances of natural resources to a state of pollution and barrenness. "Adam's Rib" is the name of a transformed space that has been degraded. As Iovino indicates, "the combination of toxic substances and 'toxic' practices produces toxic places and toxic bodies" (2012, 450). All of these toxic landscapes, then, are "storied matters" that tell narratives tracing EuroAmerican modernization and globalization that have transformed and have degraded Native Americans' homelands.

Hogan's fiction is corroborated by nonfictional scientific and health studies. According to environmental health experts L. J. S. Tsuji et al. (2005) there are many abandoned radar station sites that were built in the 1950s in the Arctic regions of North America, mainly in Canada. These are "sources of contamination by polychlorinated biphenyls" (Tsuji et al. 2005, 903). The sites are located in the western James Bay region of northern Ontario, Canada, on the Albany River. People from that region "still harvest and consume traditional wild meats, berries, and plants" (Tsuji et al. 2005, 904). The result of a scientific analysis of the region suggests that contaminants originating from the site are associated with "body burdens in Fort Albany" (Tsuji et al. 2005, 908). Toxins that are invisible to the human eye are storied matters that operate at multiple scales in both human and fish bodies. Hogan's novel represents how toxins affect minority and Indigenous communities at scales that span the microcosmic effect on individual bodies to the macrocosmic effect on communities (Adamson 2012b, 41).

In *Solar Storms*, as Angel and her grandmothers travel back to the North, they see how the massive hydroelectric project is altering the landscape: "The land was being drilled to see what else could be taken, looted, and mined before the waters covered this little length of earth" (Hogan 1995, 219) and the project will "guide the waters, narrow them down into the thin black electrical wires that traversed the world. They [the powerful] wanted to control water, the rise and fall of it, the direction of its ancient life. They wanted its power" (Hogan 1995, 268). Hogan's story forecasts the consequences that the Inuit are suffering today. As the Inuit hunter and sculptor Simeonie Kavik comments in *People of a Feather*, "The ice is harder to understand now. Hunting is more dangerous" (44:16–44:23). This comment illustrates why resistance to EuroAmerican ideologies that degrade relationships between local peoples, animals, plants, waters, and lands is materially vital to Native American oral narrations and to novelists such as Linda Hogan.

Stories Matter

Native American oral traditions have a long epistemology that show that the Earth is a multiverse, shared by fauna and flora and other living and nonliving forms at multiple scales. Seeing the construction of dams as violence against the circulation of water on the planet and as damage to ecosystems, animals, and people, *Solar Storms* fictionally introduces the notion of what I am calling a "network of networks" that takes the form of activism in which tribal people attempt to unravel the ideology of globalization from "below" (Appadurai 2001b, 3). In the novel, this "network of networks" is based on stories that Angel has learned from her grandmothers on their journey north. These stories teach her about "matters," or "storied matter" that reveal information about herself, her community, and the waterscape that her people rely on for their sustenance and survival.

In the novel, Angel was taken away from her abusive mother as a baby and reared in different foster families in Oklahoma. She returns to her birth place and journeys with her (grand)mothers to the northern ancestral homeland to find her birth mother, Hannah Wing. The long canoe journey makes Angel *re*member, or re-map, her birth place and waterscapes through the eyes of her grandmothers. As Hogan portrays in the novel, there is memory in the cells, memory in the brain, that connects them to the region. What Angel discovers is a nature-culture continuum. Stories that Angel hears on the journey are part of a world, she says, that "only my body remembered" (Hogan 1995, 79). When Angel and her grandmothers finally arrive at the northern town, traveling through the waterscapes, Angel senses: "A part of me remembered this world, as did all of Dora-Rouge; it seemed to embody us" (228). Angel's body responds to the waterscape: "It was as if it were inside me already, the future, alongside a memory of place, people, and even hardship" (137).

For Indigenous people, stories of "people, animals, and the land" are endowed with historical memory (Hogan 1995, 122). In Native American cultures, "history" is not objective knowledge one derives from textbooks. Rather, in Indigenous cosmologies, the world is made up of multiple dimensions and interactive relationships between humans and other species, nonhumans, and the natural world, which can be understood as multispecies ethnography. "Storied matters," or how different human and nonhumans communicate with each other and live in relationship create alternative realities. And as Joni Adamson (2012b, 40) explains, "multispecies ethnography calls upon humans to radically rethink categories of analysis as they pertain to life at multiple scales," from the microcosmic to the macrocosmic. This is why Angel argues at the end of the novel, "We need our own story, they have the wrong story on us" (Hogan 1995, 302). The elders' duties are to keep the stories, with their alternate realities, alive.

Uncomfortable with the narratives she learned as a marginalized foster child, Angel finds her grandmothers' stories "call me home" (Hogan 1995, 48). At the beginning of her journey, when Angel feels trapped in her self-indulgent world and is still not used to the life on Fur Island with her grandmother Bush, she decides to leave. But, she finds herself drawn to Bush's stories of her mother, who is represented by the animal wolverine, and the history of the Fur Island. Day by day, as Angel hears more stories about herself, her mother, and the past, she begins to accept and see these alternate realities, or multispecies worlds: "I began to like them, and the stones, and the still-open windows. I grew accustomed to the green reaches of the vines and the floors that creaked at night when Bush walked over them" (77). Hers is a dawning multispecies imagination: "I lived inside water. There was no separation between us. I knew in a moment what water was ... In that moment I understood I was part of the same equation as birds and rain" (78–79).

Storied matters help restore Angel's relationship with her community, the land, and animals and plants. At Bush's house, instead of pushing the *lianas* (a kind of long-stemmed vine) aside and being fearful of swimming in the cold lake, Angel, learns to open herself to the natural world around her. The boundaries between exterior and interior, between inside and outside, are blurred. Dora-Rouge, Angel's great-great-grandmother, also feels her ancestral land is calling her and talks to the churning river by making a pact with it. Like her grandmothers, Angel recognizes "the voices of the world, of what was all round us—the stones, the waters flowing toward their ends, the osprey with its claws in fish, even the minnows and spawn. I heard trees with their roots holding ground" (181). For her and her people, nature/water has its own agency, outside of human control and human will. Dora-Rouge talks to the water, and under Bush's influence Angel swims in the water and embraces the world.

The multispecies imagination that Angel receives from her grandmothers enables her entry into the biosemiotic world of plants and animals. She learns that plants, animals, and nature "see" humans, and respond to their actions. On her canoe journey to the north with her grandmothers, her mind is imprinted with visions and dreams. For example, when Angel's great-grandmother Agnes is sick during the canoe journey, Angel dreams of the plant that Agnes will need. Therefore, when the four women journey to the North, Dora-Rouge even says, "Angel, you are the one who dreamed the plant. I think you should go. Maybe the plant will call to you. Maybe it would be easier for you to find" (206). Indeed, Angel feels the plants summoning her through dreams and she hears their songs (119). Dora-Rouge and Angel's way of thinking of plants is not odd at all in the contemporary study of biosemiotics. In a biosemiotic analysis of plants in the film *Avatar,* Adamson notes that "communication among plants can credibly be explained" (2012a, 156). According to Stefan Helmreich, biosemiotics is "a field of inquiry dedicated to understanding how living things perceive and interpret their environments, whether these are mediated through language, chemical gradients, or intensities of light" (187). In other words, traditional Indigenous stories about the sentience of plants, which used to be degraded as "superstition" or "myth," are being proven through biosemiotics, and scientists are showing that plants do perceive and "communicate" with other living beings around them. Although Hogan is referencing traditional medicines and ethnobotanical knowledge, Angel's dream of plants could also be said to be pointing towards these new understandings of biosemiotic signaling within the plant world.

Hogan's depiction of Angel's growing Indigenous scientific literacy shows Angel to be cultivating an understanding of an alternative cosmos, a cosmos that might be described as "the unknown constituted by multiple, divergent worlds" (Stenger 2005, 995).[3] For example, Husk, a tribal man who helps Bush on the Fur Island, takes "Indigenous scientific literacy" as a proof of Indigenous thinking that is just as sophisticated and meaningful as Western science. Husk is the one who studies science to back up his tribal knowledge, which regards the world as "a dense soup of love, creation all around us, full and intelligent ... Even the tools and the fishhooks [are] alive" (Hogan 1995, 81). It is a multiverse that he sees and that Indigenous traditions often see represented in dreams. He indicates that "maybe visions, dreams, or memories existed because time, as Einstein thought, was not a straight line" (120). Angel recognizes her people "could talk with spirits, could hear land speak" (189). As Dora-Rouge puts it, they "talked to the churning river" (193). Communication with matter or spirits uses the language that "animals and humans had in common" (170) and "long before the notion of civilization, we knew the languages of earth, water, and trees" (334). Little by little, influenced by her own tribal people, Angel senses everything around her. To become a mindful being, Angel discovers, "My eyes saw new and other things. My ears heard everything that moved beyond the walls.

I could see with my skin, touch with my eyes" (120). She keeps feeling "new senses came to me" (127) because both her mind and physical body are changing (135) in accordance with the rhythms of the water.

On her journey, Angel gains a multispecies imagination that allows her to recognize the agency of the land, the water, and the multiple species that live in these environments. Angel perceives "invisible" matter:

> I felt there were eyes around us, peering through trees and fog. Maybe it was the eyes of land and creatures regarding us, taking our measure. And listening to the night, I knew there was another horizon, beyond the one we could see. And all of it was *storied land*, land where deities walked, where people traveled, desiring to be one with infinite space. (177; emphasis mine)

This recognition of agency of land and water is enhanced by Angel when she uses nonhuman metaphors to describe her grandmothers: "Bush opened like the lilies" and "[Bush] was water" (176; 182); Agnes empathized with the bear while the bear had been kicked and abused in the cage, singing an old bear song which only very old people had heard of; "[Dora-Rouge] read things in the moving of waters" (177). All in all, "The four of us became like one animal" (177).

Angel and her grandmothers see how the dam construction results in the creation of an apocalyptic landscape for the tribal people. The water and land Angel and her grandmothers see is "broken, water-split lands" (185). "It showed a place with numerous tiny islands scattered through the water ... some of the waters leading to [Se Nay River] were now only mud ... the Big Arm River having been diverted into the Se Nay ... the river was a deafening roar" (191). Angel discovers that "everything had changed" (193). "With more than one dam being built, much land was now submerged. An entire river to the north had been flooded and drowned. Other places, once filled with water, were dry (205). Angel draws a conclusion: "The world had changed as we traveled, and in such a short time" (211).

Hogan's end-of-world waterscape is fictional but correctly anticipates the consequences being faced today in the real world as portrayed in *People of a Feather*. By explaining the reason why he produced the documentary, Joel Heath raises questions about the effects of large-scale development and its connections to climate change on sea ice ecosystems, eider ducks, and people. He shows how Arctic eider ducks are threatened with extinction because they are getting trapped by changing sea ice patterns that affect their ability to obtain enough food to survive winter. Because of insufficient *polynyas* (areas of open water persisting through winter) in the ice, the ducks are forced to use too much critical energy diving for their food. Heath decided to make the documentary after interviewing several elders in the community, who observe:

> —I've been aware that the currents around the islands ... are not as strong now due to hydro projects on the mainland.
> —Currents give us the food for all living things in the sea ... and us too.
> —When the tides turn and the currents change direction ... the floe edges clamp shut ... so the eiders have to find open water ... and that's when they fly to the polynyas. They don't stay there all winter. When the polynya starts to get small, they move.
> —Some polynyas close over but others stay open ... where the currents are strong.
> —Since the damming began our currents have not been as strong. With weaker currents the polynyas may not stay open all winter. The first question we asked is what's going to happen with the eider ducks? (45:50–47:05)

Heath's documentary shows the Inuit helping eider ducks by opening up more *polynyas* so they can obtain food more easily. The Inuit say, "If we don't help them, they're all just going to die off" (1:19:33–1:19:34). Some human activities are harming the eider ducks, while the Inuit are trying to maintain and preserve their relationship to the ducks. In other words, human beings are "the primary agents" (Adamson 2013, 181) causing a transition to the Anthropocene, or Age of the Human.

Characters in the novel and people in the film all emphasize the importance of circulating water and of understanding human-nonhuman connections. In other words, waterscapes in *Solar Storms* and *People of a Feather* are a "storied matter" that forms a "network of networks" for First Nations. Both cosmopolitics and the connections built between humans and nonhumans through canoeing oceans and rivers are strengthened and amplified in the film and the novel through the imagination of multispecies relationships.

Cosmopolitical Activism

Although climate change was not yet in the consciousness of the general public at the time that Hogan wrote *Solar Storms* (1995), the Inuit people had long been suffering its consequences, as evidenced in *People of a Feather*. The future for the Inuit is at stake: "That's the fresh water they put into the salt water in James Bay ... That's the rivers that have been dammed. When they finish building the dams, we won't have any salt water left" (42:53–43:12). Even though many of the people in the film joke about their situations, which can be understood as a kind of "survival humor,"[4] they are facing the extinction of eider ducks and seals due to the impacts on water currents from the dam.

This real-world situation is anticipated in *Solar Storms*. Fur Island, for example, "like all the other broken pieces of land, had begun to shrink. We wanted to salvage the plants, the corn, whatever could be taken away"

(Hogan 1995, 337) and "soon the water would approach Bush's house ..." (338). Habitats start to disappear. "Ammah's, where birds nested and hope roosted, simply vanished into the water. No one could account for what happened to it. But from the distance of that day, we heard the sound of rumbling machines and the rocks and trees breaking at another new site of construction" (272–73). Hogan illustrates what happens when places/waterscapes disappear due to a dam. She shows how devastation and ruin of the land is followed by displacement of people. Yet the characters of the novel face this crisis, their persecuted past, uncertain present, and endangered future with action and resistance.

In *Solar Storms*, Hogan sees "change" as necessary for the survival of Northern Indian cultures as well as other tribal cultures. The canoe journey, with its "grandmothers and grandchild" bond offers a sense of an activist spirit to mark "a *moment of transition, a moment of decisive intervention*" (Hay 1996, 253; emphasis added). Angel and her family and community are modeling a cosmopolitical resistance in the "developing" world. The novel, which emphasizes Angel's recognition of the more-than-human world (like the film with its attention to the ocean and eider ducks) not only questions presumptions about "the human" as a center but also reinforces the notion of a "cosmos of relations" or the world as a "dynamic space that includes humans and the whole community of beings that exist" (Adamson 2012b, 39). Though the whole novel is narrated mainly in Angel's voice, her grandmothers also participate in the narrative, which implies that she is not the only important person, but one voice amidst a chorus of voices of her ancestors. Angel feels an obligation to fight for her people's lands and waters as well as their culture and their multispecies community because humans and nonhumans are seen as kin. She sees a multispecies world being destroyed by the dam and states, "For my people, the only survival is resistance" (Hogan 1995, 325). A cosmopolitical resistance involves working not only for one species but for multiples species living in relationship. The ongoing displacement and devastation caused by the dams energizes Angel and her family "to fight" so that they can "respect ourselves, we Beautiful People" (325).

The Beautiful People's "grassroots organizing" (308) represents "a network of networks" that Chadwick Allen has describes as a trans-Indigenous identity (2012, 12).[5] In *Solar Storms*, Hogan presents the grassroots movement forming in resistance to the hydroelectric project, as an early example of what has been called the environmental justice movement, and more recently, cosmopolitics. The relation between "community" and "place" in the novel does not remain a one-to-one correspondence because people from other places who join the fight against the dam engender a sense of cross-Indigeneity. In order to blockade part of the dam project, people from different tribes and various professions come to assist: Arlie Caso House, an Ojibwa man, a political prisoner, and a master strategist; Lozen, a healer, a fighter, a locator of the enemy; twelve men came with Arlie, Apostles

(Hogan 1995, 308–9). As Rachel Stein observes, "Hogan represents environmental justice activism as a fundamental form of family and community solidarity" (2002, 200).

The anti-dam activism in *Solar Storms*, as a form of environmental justice activism, takes two forms. First, by refusing to consent or to sign any papers authorizing construction of the dam, to take compensation, or compromise, the people reveal by their actions that they refuse to allow their multispecies world to be destroyed. They articulate that following "white laws," which they were taught in their boarding-school textbooks, and which attempted to instill in them the "truth" of science, history, and culture according to Western philosophies, would degrade Indian cultural heritage and "kill the world" (283). They understand that Western development "didn't hold a thought for the life of water, or a regard for the land that sustained people from the beginning of time" (279). On the other hand, writing and telling their own story through their own forms of media is a way to resist violence while still taking a stand against dam construction. Angel uses the radio to tell the public how Indigenous leader Tulik's house, where the activists are organizing, is surveilled by the army (292) while white radio announcers cover up or ignore the protest. Bush and Luce use newspapers, typewriters, and cameras to document the devastation of the region and to broadcast their story (300). Media attention successfully draws outside attention. More people come to help due to Bush's stories and photos. Finally, they decide to block the railroad tracks, because even if they lose, they would have self-respect (305). These resistance actions foster stronger cohesion and solidarity.

Hogan's fictional characters represent Indigenous peoples who are organizing in the real world.[6] This is why *Solar Storms* offers insights into the cosmopolitical movement that is being documented in films like *People of a Feather*, which is representing Indigenous resistance to the human activities that are contributing to climate change. The documentary voices the dilemma that Inuit people are facing. After talking to the elders and working with the hunters for many years, Joel Heath observes that "I've learned a lot about sea ice ecosystems … and how changing the hydrological cycle can influence the marine environment. But we're only just beginning to understand these effects and we still have a lot of questions. What about the other animals that rely on sea ice? And what about the rest of the world outside of Hudson Bay? The deeper I dug, the bigger I realized the scale of the issue" (48:20–48:44).

Heath's film provides evidence of the cumulative effects of the growing complex of hydroelectric projects at James Bay. A study of this type was promised to the Inuit people but has never been provided. Indeed,

> In the early 1990's, Inuit and Cree protested with supporters in New York City and were able to stop additional hydroelectric developments. But it was just a delay. A similar project became operational in 2010, significantly expanding the hydroelectric complex. The new

> project was approved on the condition that a study of the cumulative effects on the marine ecosystem would be undertaken. To date, no such study has been done. (1:26:05–1:26:50)

Though the battle still goes on, according to the film, "The Arctic Eider Society has been formed to assist Inuit in community-based environmental monitoring programs ... to provide research and education combining Inuit and scientific knowledge on sea ice ecosystems ... and to support development of energy solutions that work with the seasons, not against them" (1:27:30–1:28:00). By practicing an activism through the production of a documentary, the Inuit are seeking to influence decisions about future construction of dams at James Bay. Heath reminds his audiences, "We're working against the seasons of our hydrological cycle. We still have a lot to learn, but we can tell you from our perspective here, ... things are getting a lot noisier, and less predictable" (49:18–49:30).

Conclusion

People of a Feather documents connections between changing sea ice, disappearing eider ducks, and the Inuit dependence on eider ducks for eggs and down for their clothing" (Heath 2004). It shows that when the eider ducks are gone, so too will be the Inuit way of life. Hogan's *Solar Storms* shows that like millions of people displaced by development projects around the world, at the end of the novel, Angel must migrate to the city (Hogan 1995, 341). She has learned the meaning of a multispecies network of networks taught to her by her grandmothers, who gave her "a language, a story, to shape [herself] by" (94). "I learned how I came from a circle of courageous women and strong men who had walls pulled down straight in front of them until the circle closed ... but some lived, some survived this narrowing circle of life" (107).

Hogan's fiction calls for respect for Native American cultures, esteem for their multispecies worldviews, and a retelling of history from a Native American perspective. Through the reconnection and resistance shown in the novel, Hogan offers self-respect and self-empowerment to her own people and a new understanding of Native American cultures to non-Native American readers. Hogan's stance does not reject dominant culture but asks if it is worthwhile to seek human well-being at the cost of annihilation of multispecies worlds. Likewise, the documentary *People of a Feather* advocates for more solutions that "work with the seasons," the environment, and the ecosystems. Both the novel and the film can be analyzed, through material ecocriticism, not just as artistic speculation, but as clear-sighted explorations of social injustices, displaced humans, accelerating environmental degradation, changing waterscapes, and dying wildlife, and at the same time, of the possibility of building a network of networks to resist destructive forces, and fight for the protection of a cosmos of relations.

Acknowledgments

This chapter was supported by a grant from the Ministry of Science and Technology of the Republic of China (under the grant number MOST 103–2410-H-211–003-). I am also very grateful to Professors Joni Admason and Salma Monani for their insightful and valuable comments on various versions of this chapter.

Notes

1. "A network of networks," as the chapter further explains, is a term I use to draw together material ecocritical actor-network theories (Latour, 2013), globalization from above and below (Appadurai 2001a), and cross-Indigeneity at every scale from the cosmic to the microcosmic (see especially Adamson 2012b).
2. For example, Laura Virginia Castor's (2006) reading of the novel emphasizes Native American views of landscape and distinctive place analysis as relevant to each character. Reading from the eco-feminist point of view, Silvia Schultermandl reads the novel as a resistance story to counter "the dislocation of the individual from her natural and cultural landscape" (2005, 67). Ellen L. Arnold (2004) points out the theme of "beginnings" and "origins," which plays an important role in Native American Literature such as in *Solar Storms,* to highlight one of the spiritual tenets in the novel: "Our beginnings were intricately bound up in the history of the land" (Hogan 1995, 96). Ann Fisher-Wirth (2003), Roseanne Hoefel (2001), and Jim Tarter (2000) all regard Hogan's landscape and cosmology in the narrative as "storied Earth" and "dreams of Earth" to emphasize that Native American oral storytelling cannot be separated from its geographical locations and its actual physical places. In particular, Tarter draws attention to the issue of environmental justice to give weight to the combination of spiritual journey and political resistance. German ecocritic Christa Grewe-Volpp (2002) extrapolates the image of the ecological Indian and indicates the possible pitfalls raised by Noël Sturgeon (1997) when approaching the novel from the perspective of ecofeminism.
3. Marisol de la Cadena (2010), Isabelle Stengers (2005), and Deborah Bird Rose (n.d.) all talk about multispecies worlds.
4. The term "survival humor" is used often in discussions of Native American's humor. Louise Erdrich even regards "survival humor" as "essential to Native Americans' response to their overwhelming experience of cultural catastrophe ..." (Stookey 1999, 20).
5. Hogan speculates on the realities of the boundary struggles. These realities play out in *People of a Feather*. As an activist documentary, *People of a Feather*'s political bias must be recognized, however. As its accompanying website (http://www.arcticeider.com/en/about) shows, native struggles are grounded in cooperative action. See, for example, this passage from the website: "Recently, we've formed the East Hudson Bay Network, expanding our community-based programs from Sanikiluaq, Nunavut to include Inukjuaq, Umiuaq, Kuujjuaraapik and Chisasibi in northern Quebec, creating the capacity for large scale coordinated efforts to begin assessing cumulative impacts. Working closely with the Hudson Bay Inland Sea Initiative we are also establishing a research consortium

and governance structure for Hudson Bay, necessary to overcome the inter-jurisdictional challenges that have held back progress on addressing cumulative impacts and Inuit concerns to date" (see The Arctic Eider Society website: https://arcticeider.com/en/about).

6. The James Bay protests have being reported in news articles. For example, the *Montreal Gazette* (Nov. 29, 1973) showed how Indian people on the Boundary Waters between Minnesota and Canada had been devastated because of the construction of the dam. (https://news.google.com/newspapers?nid=1946&dat=19731129&id=UpMuAAAAIBAJ&sjid=XKEFAAAAIBAJ&pg=4578,3447895&hl=zh-TW).

References

Adamson, Joni. 2012a. "Indigenous Literatures, Multinaturalism, and *Avatar*: The Emergence of Indigenous Cosmopolitics." *American Literary History* 24 (1): 143–62.

———. 2012b. "Whale as Cosmos: Multispecies Ethnography and Contemporary Indigenous Cosmopolitics." *Revista Canaria de Estudios Ingleses* 64 (1): 29–45.

———. 2013. "Environmental Justice, Cosmopolitics, and Climate Change." *The Cambridge Companion to Literature and the Environment*, ed. Louise Westling, 169–83. Cambridge, UK: Cambridge University Press.

Allen, Chadwick. 2012. "A Transnational Native American Studies? Why Not Studies that are Trans-Indigenous?" *Journal of Transnational American Studies* 4 (1): 1–22.

Appadurai, Arjun. 2001a. "Grassroots Globalization and the Research Imagination." In *Globalization*, ed. Arjun Appadurai, 1–21. London: Duke University Press.

———. 2001b. *Globalization*. London: Duke University Press.

Arnold, Ellen L. 2004. "Beginnings Are Everything: The Quest for Origins in Linda Hogan's *Solar Storms*." In *Things of the Spirit: Women Writers Constructing Spirituality*, ed. Kristina K. Groover. Notre Dame, IL: University of Notre Dame Press.

Castor, Laura Virginia. 2006. "Claiming Place in Wor(l)ds: Linda Hogan's *Solar Storms*." *Multi-Ethnic Literature of the United States [MELUS]*, 31 (2): 157–80.

Coppinger, Raymond, and Will Ryan. 1999. "James Bay: Environmental Considerations for Building Large Hydroelectric Dams and Reservoirs in Quebec." In *Social and Environmental Impacts of the James Bay Hydroelectric Project*, ed. James F. Hornig, 41–72. Montreal: McGill-Queen's University Press.

de la Cadena, Marisol. 2010. "Indigenous Cosmopolitics in the Andes: Conceptual Reflections beyond 'Politics.'" *Cultural Anthropology* 25 (2): 334–70.

Fisher-Wirth, Ann. 2003. "Storied Earth, Storied Lives: Linda Hogan's *Solar Storms* and Rick Bass's *The Sky, the Stars, the Wilderness*." In *From the Center of Tradition: Critical Perspectives on Linda Hogan*, ed. Barbara J. Cook, 53–66. Boulder: University of Colorado Press.

Grewe-Volpp, Christa. 2002. "The Ecological Indian vs. the Spiritually Corrupt White Man: The Function of Ethnocentric Notions in Linda Hogan's *Solar Storms*." *Amerikastudien / American Studies* 47 (2): 269–83.

Grinde, Donald A., and Bruce E. Johansen. 1995. "Native American Environmental Testimonies: The Last Frontiers of Ecocide." In *Ecocide of Native America: Environmental Destruction of Indian Lands and Peoples*, ed. Donald A. Grinde, Bruce E. Johansen, and Howard Zinn, 221–60. Santa Fe, NM: Clear Light Publishers.

Hay, Collin. 1996. "Narrating Crisis: The Discursive Construction of the 'Winter of Discontent.'" *Sociology*. 30 (2): 253–77.

Helmreich, Stefan. 2009. *Alien Ocean: Anthropological Voyages in Microbial Seas*. Berkeley: University of California Press.

Hoefel, Roseanne. 2001. "Narrative Choreography toward a New Cosmogony: The Medicine Way in Linda Hogan's Novel *Solar Storms*." *FEMSPEC* 2 (2): 33–47.

Hogan, Linda. 1995. *Solar Storms*. New York: Simon & Schuster.

Hornig, James F., ed. 1999. *Social and Environmental Impacts of the James Bay Hydroelectric Project*. Montreal: McGill-Queen's University Press.

Iovino, Serenella. 2012. "Stories from the Thick of Things: Introducing Material Ecocriticism." *Interdisciplinary Studies in Literature and Environment [ISLE]* 19 (3): 449–60.

Iovino, Serenella, and Serpil Oppermann. 2012. "Theorizing Material Ecocriticism: A Diptych." *Interdisciplinary Studies in Literature and Environment [ISLE]* 19 (3): 448–75.

Latour, Bruno. 2013. *An Inquiry into Modes of Existence: An Anthropology of the Moderns*. Trans. Catherine Porter. Cambridge, MA: Harvard University Press.

Oppermann, Serpil. November 2013. "Material Ecocriticism and the Creativity of Storied Matter." *Frame: Journal of Literary Studies Special Issue on Ecocriticism*. 26 (2). 57–72.

People of a Feather. 2011. Directed by Joel Heath. Montreal, Quebec: Sanikiluaq Running Pictures Ltd. DVD.

Rose, Deborah B[ird]. (2012). "Multispecies Knots of Ethical Time." *Environmental Philosophy* 9 (1): 127–40.

Roy, Arundhati. 1999. "The Greater Common Good." *Frontline*. November 16. Magazine: cover essay. Accessed May 20, 2015. http://www.frontline.in/static/html/fl1611/index.htm.

Schultermandl, Silvia. 2005. "Fighting for the Mother/Land: An Ecofeminist Reading of Linda Hogan's *Solar Storms*." *Studies in American Indian Literature* 17 (3): 67–134.

Shiva, Vandana. 2000. *Water Wars: Privatization, Pollution and Profit*. Cambridge, MA: South End Press.

Stein, Rachel. 2002. "Activism as Affirmation: Gender and Environmental Justice in Linda Hogan's *Solar Storms* and Barbara Neely's *Blanche Cleans Up*." In *The Environmental Justice Reader: Politics, Poetics, and Pedagogy*, ed. Joni Adamson, Mei Mei Evans, and Rachel Stein, 193–211. Tucson: The University of Arizona Press.

Stengers. Isabelle. 2005. "The Cosmopolitical Proposal." In *Making Things Public: Atmospheres of Democracy*, ed. Bruno Latour and Peter Weibel, 994–1004. Cambridge, MA: MIT Press.

Stookey, Lorena Laura. 1999. *Louise Erdrich: A Critical Companion*. Westport, CT: Greenwood Press.

Sturgeon, Noël. *Ecofeminist Natures: Race, Gender, Feminist Theory and Political Action*. New York: Routledge, 1997.

Tarter, Jim. 2000. "'Dreams of Earth': Place, Multiethnicity, and Environmental Justice in Linda Hogan's *Solar Storms*." In *Reading Under the Sign of Nature: New Essays in Ecocriticism*, ed. John Tallmadge, 12847. Salt Lake City: University of Utah Press.

Tsuji, L. J. S., B. C. Wainman, I. D. Martin, J.-P. Weber, C. Sutherland, and E. Nieboer. 2005. "Elevated Levels of PCBs in First Nation Communities of the Western James Bay Region of Northern Ontario, Canada: The Use of Correspondence Analysis to Identify Source of Exposure." *Bulletin of Environmental Contamination and Toxicology* 75 (5): 903–9.

Warner, Stanley, and Raymond Coppinger. 1999. "Hydroelectric Power Development at James Bay: Establishing a Frame of Reference." In *Social and Environmental Impacts of the James Bay Hydroelectric Project*, edited by James F. Hornig, 19–38. Montreal, Quebec: McGill-Queen's University Press.

10 *tinai*-Documentation as Ecocultural Ethnography

My Experience with the Mudugar

Rayson K. Alex

Tribal people are the most oppressed traditional cultural communities in India. Deforestation, land grabbing, linguistic discrimination in schools and government offices, lack of medical facilities, religious conversions, enforcement of modern agricultural practices, drastic change of food patterns, lack of political representation, spread of genetic diseases, sexual violence on women, sand and rock mining, and acute poverty are central to tribal oppression and social exclusion. There have been sporadic issue-based reports on tribal people in the media and studies by academics, such as newspaper articles on the bauxite mining in Andhra Pradesh (Sudhir 2015), low literacy rates and poverty among tribal people ("Low Literacy" 2014), tribal development and prostitution ("Tribal Women" 2013) and diseases of tribal people in Andaman Island (Vidal 2009), to mention a few. However, detailed ethnographic studies focusing on eroding tribal life-ways have not been adequately done in India. Even as this chapter is being written, India's aggressive neoliberal developmental schemes are particularly threatening to Indigenous tribal communities, pushing them off their subsistence and communal lands. The passing of the recent Mines and Minerals (Development and Regulation) Amendment Bill 2015 is one such example of developmental policy in India. The bill does not have any provisions for safeguarding the rights of the original owners of the land (Karat 2015). However, this issue of environmental justice has not been adequately discussed in the media or among academics. Therefore, any cultural documentation of the people gains immense social and political relevance. Such documentation can play a vital role in revising centuries of feudal, colonial, and now neocolonial power dynamics that discriminate against Indigenous communities and their ways of knowing and living.

This chapter's focus is on the Mudugar, a tribal community living in Attappady, in the Palakkad District of Kerala State, Southern India. There are two other tribal communities, the Kurumbar and the Irular, living in Attappady. While the Kurumbar are early settlers, the Irular are more recent immigrants (Ayyappan 2012). Attappady is a mountain-valley west of the city of Coimbatore, Tamil Nadu, and is flanked by the Nilgiri mountains in the north. It is 180 kilometers by road from Coimbatore, the largest city in the area, and 80 kilometers from Palakkad, a smaller city on the border of Kerala. Because of the region's border status, the Mudugar people, along with the other tribal communities, are in constant contact with migrants from different parts of Kerala and Tamil Nadu. These relations have often been oppressive, with the Mudugar ceding territory and shifting their settlements further into the forests of the region (Ayyappan 2012); here too their livelihoods are marginalized by India's class, caste, and settler society (Kabeer 2012, 90).

In negotiating their discriminated place in society, many Mudugar are multilingual. Most of them speak Malayalam (the official language of the State of Kerala) and Tamil (the official language of the State of Tamil Nadu), along with the Mudugar *bhaasha* (literally translated, "the Mudugar language"). Because of its semiotic similarities to Tamil, I believe that Mudugar *bhaasha* is a variant of ancient Tamil, but without a script. If language and cultural patterns are criteria for understanding the origin of a community, the Mudugar, like other Dravidian communities in present-day India, are ancestrally linked to India's earliest human inhabitants. Central to their contemporary negotiations and the continued use of their language are attempts to keep alive traditions that articulate their long existence in the area, which in turn is enmeshed in a sense of deep ecocultural rootedness.

Like with all Indigenous peoples, this ecocultural rootedness cannot be essentialized, as it is tied to the long process of learning to live with the land. I have been closely interacting with the Mudugar community for the past ten years. I draw on four stories from my own process of learning about these traditions. I argue that these traditions embrace the conceptual framework of *tinai*. *tinai* is a geo-poetical convention of the early Tamil people. K. Sivathamby (1974) in his essay, "Early South Indian Society: The Tinai Concept," argues that the poetic convention is based on the social and economic characteristics of the communities in the five landscapes—mountainous, pastoral, riverine, littoral, and desert. Nirmal Selvamony builds on this proposition and identifies the social, cultural, religious, and natural features of the five landscapes (Selvamony 1990, 217–21), defining *tinai* as a "human-nature-spirit nexus" (Selvamony 2007, xv). Ecocritic Scott Slovic mentions the relevance of *tinai* as an "Indigenous notion" that is unique and could be relevant in understanding culture-specific environmental problems (Slovic 2010). According to ecocritic Annie Ingram, *tinai* "exceeds the locational specificity of "environment" and extends the "house" or "habitat" connotations of *oikos* to include notions of eco-regions, and Indigeneity (Ingram 2013). In her discussion, she points to the "sense of connectedness" that *tinai* as a concept proposes. David Damrosch (2013) acknowledges *tinai* as an "ecological view of language/literature" that highlights a "biome-based social order integrating land, man, and god (2013). Apart from the considerable amount of scholarship on *tinai* available in India, it has been identified as a useful ecocritical tool to understand ecological crises not only in a local but also in a global sense (Alex 2014).

While *tinai* is a concept with origins in South Indian contexts, it demonstrates a close resemblance to the idea of Indigenous cosmovisions, as it speaks similarly to a worldview of natureculture continuity, connecting land, memories and myths, or nature, culture, and the sacred (de la Casena 2010; Poirier 2008). The stories I outline below illuminate the traditional and ancestral cosmovisions of the Mudugar people. At the same time, these stories articulate a politics of Indigeniety. Specifically, the first story highlights the tensions of centuries of migrant influx into the Mudugar community and demonstrates how ecological rootedness is one defense against such incursions. The second story articulates this ecocultural rootedness by engaging the Mudugar relations to *paan* ("betel nut"). This rootedness enlivens the natural world with a non-hierarchical agency, which nonetheless must be carefully negotiated through attention to cosmopolitics to extend past memories into present (and hopefully future) actions that foster sustainability. The third story considers how relations with bees suggest the same understandings. The final story further expands these understandings of ecocultural rootedness by turning to the Mudugar origin story, which animates not just animals and plants but a local rock hill with sacred significance. Such engagements, I argue, demonstrate a Mudugar value of human and nonhuman entanglements that speak for a sustainable relationship and reverence with their environments. Their continued practice in contemporary times also serve as modes of resistance against long legacies of marginalization. In the face of a current onslaught of rapid, capitalist-driven development in India, not paying attention to the Mudugar and the knowledge of the land that they live on is inherently a grave environmental injustice. *tinai*-documentation attempts to address this injustice.

Ecocultural Rootedness

It was on a sunny day in February 2007, when Ayyappan, a distinguished local artist who is known in his community as Maaman ("Uncle"), my three other friends (U. A. Manoj, Arun Bose, and Radhakrishnan), and I decided to explore Attappady with our video camera. Our intention was to make a video-documentary (the edited video is titled *The Story of Mudugar* [2008]) as part of my research work for my Ph.D. in ecocriticism. An excerpt from the conversation that we had with Maaman and the edited version of which appeared in the video-documentary is given below:

"What kind of grass is this, Maaman?," I asked.
"This is the kind of grass that we use to thatch our houses," said Maaman.
[*Maamam became excited, raising his voice in a dramatic way. Stepping off the mud-road (through which we were walking down the hill) onto the slope of the green grassland, he began humming a song and then singing*

it aloud. With the sickle in his hand, he cut a handful of grass. He held the grass high above his head with both hands, holding the tips of the grass, and the sickle in his right hand. Making sure that he was within the frame of the camera, he eagerly and quite expressively remarked:]

"We tie the grass together and hrrrrrrrrrm (*produces a guttural sound*) the man and the woman are inside the house."

My friend Arun Bose (one of the documentarian team members) asked, "is the grass used for thatching?"

"Yes, yes, for thatching," said Maaman.

[*Interrupted, he continues with a different narrative.*]

"When I build a big house, my wife and I will occupy one room, my children the other, and their uncle and aunt another."

"There is no room for anyone else. If there are other people, no one in the house can sleep, if there are other people ..."

(*The Story of Mudugar* 2008)

On our way up and down the Ommala hill, the then 75- to 80-year-old Maaman engaged us with stories of the plants, animals, and birds that we saw on the way. I have known Maaman and his family since 2003. Contrary to his statement about "no room for other people in his home", he had hosted me in his house a couple of times and provided me with food and security; I have looked after his little five-year-old daughter; he has guided me through other hamlets and introduced me to other community elders. The "no room" statement that suggested Maaman's house was only for his family made me restless. After I had shared a cordial relationship with the family, how could Maaman say this to me? I wondered.

While the general reference to "other people," as expressed by Maaman, implicitly points to me and the documentarians in my group, the reference makes most sense when understood in the context of the constant influx of migrants to the Attappady region; while I am not a migrant, I am also not from the area. The Mudugar have a long history as the victims of oppressive relations with migrants. The Irular, another Indigenous community, are believed to be the earliest migrants who began wet and monocultural agriculture in Attappady. They are believed to have migrated from Coimbatore in the sixteenth century when there was a water scarcity (Velluva 1999). Migrating into the region, they clashed with the native Mudugar and Kurumbar. The Mudugar had to shift their settlements a few times in fear of the Irular and their exploitation. This explains why the Mudugar give brides to Kurumbar bridegrooms but detest any relationships with the Irular (Ayyappan 2012; Feroze 2011; "Major Tribes" 2016). Irular now are the largest tribal population in Attappady[1] (Feroze 2011). More recent Tamil-and Malayalam-speaking settlers from different parts of Kerala and Tamil Nadu continue to settle in Attappady. In the last twenty to thirty years rapid development *sans* cultural considerations have drastically altered the face of Attappady—roads have been built, education institutions established, local governmental administrative bodies established, concrete houses built, trees chopped down for development. The Mudugar have constantly had to cede their territories to settlers, and in the affront of the land loss, Maaman's words can be read not as simply a personal slight to me but as a deeply political statement—as a means of articulating his deep connection with his place, home, culture, and the community. For Maaman, his house, made from the materials of the land he toiled on and in the tradition of his people, is an identification with his ecocultural rootedness.

As much as this identification is materially grounded, it is also familial. Maaman's elder daughter is married and lives with her husband's family in Karuvera hamlet. It is Maaman's and his wife's ardent wish that their daughter and her family lived with them. The house referred to could be seen as "home," where all the members of the family live together. Home to Maaman is beyond the building structure and extends to include relationships between people and, of course, the values connected with it. When Maaman says "When I build a big house …" the "I" should not be misunderstood as an individualistic word but as having a meaning more like the pronoun "my." It can be considered identification with the community, its beliefs, and activities. In my ethnographic research, I saw that the "I" of the Mudugar is a complex unit comprising their immediate environment, their kin, "nonhuman living beings" (Naess 1995a, 226), and nonliving beings. The people in the hamlet own their land/hamlet collectively and not individually. Moreover, all activities related to the land—be it construction of a house, conducting a festival, or cultivation—are always community-based. In the present moment, the influx of migrants to Attappady and increased interference of government-encouraged capitalist development schemes stand in direct conflict with the Mudugar's philosophy of living.

Thus, even as it manifests this rootedness, Maaman's understanding of house also captures dynamism in his Indigenous thinking in the face of a changing world. In contrast to his present house, Maaman dreams of a new, big house. Maaman's aspirations to build a big house would go against the existent philosophy of the tribal community—smallness of scale, which adheres to ecological awareness and maintaining community equity. Yet, despite wanting a big house, Maaman describes that the house's structure will be similar to the rest of the houses in the hamlet, which have a uniform building structure. In Maaman's placing me as an outsider to his house, one must understand the larger context of everyday struggles of the community negotiating the changing politics of its world. Maaman draws on a sense of deep ecocultural rootedness to affirm his connection, through his house and community, to resist intrusion by those who have constantly marginalized these connections. To me, this deep cultural rootedness, which integrates the Mudugar community, their culture, and understandings of land, opened a new awareness to the political enactment of *tinai* by the Mudugar.

Traditionality, a Characteristic of *tinai's* Ecocultural Rootedness

I narrate another instance that I faced in the initial years of my research with the Mudugar community, back in 2004. My acquaintance with Maaman's family was a few months old. Maaman decided to take me to his relative's house in the same hamlet, Thunduur. On a rainy day, we started our journey to Chelli's house. To reach the house, we had to walk nearly one kilometer from the tarred road, crossing a bridge made of two logs of areca-nut poles over a stream, through the ridge of a cultivated field, to finally reach a mud road that bridged low-lying land and a hill. The slope up the hill was approximately 55 degrees. Climbing up the steep slope was an arduous task for me, as I was wearing a pair of worn-out rubber flip-flops. Twice I reached the top of the hill, only to slide down in less time than it had taken to climb up. Maaman stood at the foot of the hill enjoying the scene. The second time I slid down, he disappeared. In a minute or two he returned with a log and gave it to me for support. The then seventy-year-old man was behind me to watch my step while I grappled my way up the hill with the stick. He had asked me to leave my footwear behind, and as he hopped up the boggy ground, he picked up my footwear on his way.

Chelli's hut was visible as we reached the hilltop. Though I had noticed an approximately fifty-year-old woman sitting on the steps of her unroofed veranda, I did not know that we were going to that house. As soon as she saw Maaman, she began a conversation with him in the Mudugar language, which was audible to the entire village. Three women, with excitement and warmth on their faces, came out of their houses to see him. On his way to Chelli's house, Maaman flirted and made fun with all those who came to meet him; I followed him. When we entered Chelli's house, the others left us. Leaving us standing on the veranda, Chelli went in to get a straw-mat. Spreading the straw-mat on the floor of the veranda, she invited us to sit while Maaman asked her if she had some betel. She prepared three, keeping one for herself, giving one to Maaman, and offering me the other. That was the first time someone had offered me betel. Though I was a bit hesitant, I accepted the offer as Maaman shook his head to endorse the offer. With a smile, stretching out my hand, I took the *paan* and began chewing. Chelli's face lit up. After a minute or two, my mouth began burning as she had applied more lime than usual (as I was later told). Meanwhile, Chelli sang songs and narrated the beliefs around them. Even after I spat out the betel juice, my mouth continued to burn. After two hours or so, we bid farewell to Chelli and went down the hill. Reaching the Ashramam road (as it is called), near the Thunduur hamlet, I parted and went to Goolikadavu (a small town in Attappady) to meet my documentarian colleague, Manoj.

I narrated the story to him and we went straight to a small eatery to rinse my mouth. My mouth was sore and started burning more when I rinsed it. At the time, I made a resolution—to never consume *paan* again. However, I willfully broke my resolution a few other times, in the same hamlet when *paan* was offered to me by the community elders. The reasons for accepting and chewing paan are important to illuminating my attempts at *tinai*-realization. I was trying to understand the cultural symbolism of Chelli offering me *paan*. Why did she do this, despite the fact I was not one among them?

To understand the value of *paan* to the Mudugar, I would like to quote two excerpts, one from *Kuruntokai* (song 65, lines 10–14), a Tamil Sangam song, and the other from Kapilar's *Purananuru 107* (quoted in Bate 2009). These songs from the prehistoric era are quoted to explain the depth of ecology in Tamil Indigenous cultures by considering the example of *paan*. *Paan* leaf and the act of chewing it have multiple cultural implications. F. M. Franco quotes Nirmal Selvamony's translation of *Kuruntokai,* the Tamil Sangam song, as follows:

> Peering, blabbering and slightly bowing, he quipped,
> "Who are you that waits here at an untimely hour?"
> "Saying so, without moving away from me, like a bull
> that chanced upon hay,
> he untied his bag and said to me,
> 'O, girl, will you have some betel?
> Come on, have some!'"
>
> (Franco and Narasimhan 2008, 53)

In Kapilar's *Purananuru 107,* there is a mention of the metaphor of the "red-tongued poets" (*chennapulavar: chen* = red; *pulavar* = poet). An excerpt of the poem is given below:

> Pari! Pari! so many
> red-tongued poets praise
> this king of singular excellence
>
> (Bate 2009, 38)

Bernard Bate, author of *Tamil Oratory and the Dravidian Aesthetic: Democratic Practice in South India*, relies on George Hart and A. K. Ramanujan's interpretations of the metaphor of red. Red is translated as "fine," "cultivated," "good," "skillful," "beautiful" and "eloquent" (quoted in Bate 2009, 38). *Chentamil* (literally, "red Tamil") means "beautiful Tamil." Literally, red refers to the "betel-stained mouths" (Bate 2009, 38) of the poets. Merlin Franco and D. Narasimhan, in their essay "Ecological Elements in the Songs of the Poraja and the Ancient Tamils," explain the context of the Tamil song. A Brahmin who sees a girl attempts to get closer to her and offers her *tampulam* (in the Tamil tradition, *paan* is called *tampulam*), with the motive of seducing her while she awaits her lover (Franco and Narasimhan 2008, 55).

Offering *tampulam* is a practice of the Dravidian families and irrespective of caste and gender is used as a sign of respect or greeting. Despite the Brahmin being a stranger to the girl, she does not consider it offensive when the Brahmin offers *tampulam* to her. She only finds his intentions abominable. The association of the color red with pride, goodness, skill, and art refers to the status of poets in the King's court. But as chewing *paan* is the tradition of the Tamils, elders belonging to all castes chew it. Chewing *paan* also signifies dignity, respect, and goodness. Tamil traditions and early Tamil literary texts reiterate the practice of chewing *paan* as rooted in culture.

My subsequent engagements with the Mudugar suggested that the cultural roots of their *paan* traditions are also deeply ecological. For example, in keeping with the principles of *tinai*, *Pariyappanam* is a *paan* ritual practiced among the Mudugar to negotiate harmonious cosmopolitical relations with the multispirited worlds in which they live. Sylvie Poirier describes these negotiations as "constantly being embodied, actualized, or (re)enacted ... One's ancestral (intimate) connections intrinsic to the relational self are acquired through conception, inheritance, rituals, or dreams relations, depending on local tradition" (2008, 79–80). *Pariyappanam* is the return of the *penpanam*—the money a person had received or given to the elders of the hamlet during a ritual-like birth or marriage. It is a ritual of Mudugar that fits Poirier's notion of *ancestrality*. The members of the family settle all liabilities of the dead person during this *pariyappanam*. The dead body is kept at the center of the hamlet. The central ground of the hamlet is used for various other purposes, like drying clothes, cereals, or cow dung, and for conducting games, dances, and rituals. During the death ceremony, some members (men, women, and children) of the hamlet dance around the dead body singing songs. Some others, especially the elders (only men) and the hamlet officials, gather in a shaded corner of the central ground to carry out *pariyappanam*. All the well-wishers, friends, and relatives of the dead person contribute to the *pariyappanam* collected by the chieftain on behalf of the dead person's family and the hamlet. The money that is handed over to the chieftain (usually 1, 10, 50, or even 100 rupees) is wrapped in a betel leaf, with an areca nut (Sakthivel 2005, 51).

The leaves will later be used to prepare *paan,* while the money will be used for the collective good of the family or the hamlet, for such things as house renovations, expenses of the ritual, making country liquor for the hamlet, and agriculture. "Ancestrality (or traditionality) is not to be understood here solely as a reference to the past, a mythical past, or a genealogy, but rather as the ongoing expression of agencies and persons who were here before and who left some things behind (substances, signs, powers, or memories) that continue to participate in the unfolding of the world" (Poirier 2008, 79). The play of ancestrality and traditionality in the cosmos of Mudugar brings the past to life (through experiences, memories and beliefs), thereby extending the cosmos to the present and the future—as a circular time frame.

The traditionality in the use of *paan* demonstrates what I call *eco-culture specificity* in the performance of the ritual, *pariyappanam*. The betel plant (*Piper betle)* is believed to have originated in Asia (Rooney 2014) and I have seen many Mudugar houses growing betel plants in their backyards. The Mudugar claim that the plant is native to the place. Transformed into a sacred entity in the *pariyappanam* ritual, the betel takes on the status of an ecocultural keystone. Ecologically, a keystone species is considered the most important species in an ecosystem, which is capable of regulating the entire ecosystem (Visser, Thébault, and de Ruiter 2012, 60). Similarly, we can think of an ecocultural keystone (a living or nonliving entity) as essential, as it has ecological and cultural implications for an entire community (both humans and their extended environment). Another instance will elaborate this point further.

Ecocultural Rootedness as Cosmopolitical Negotiations

Cheeru is a ritual performed by Mudugar once in twelve years much like the *Kurinji* flowers (*Strobilanthes kunthiana)* that bloom once every twelve years in the Nilgiris (*niila* = blue; *giri* = mountain), in the Western Ghats. *Cheeru* is a ceremony celebrated by the Mudugar to respect parents and others who have died in the span of the previous twelve years. This unique ritual is carried out at the end of every twelve years. The skulls of the dead bodies are dug out and transferred safely to a cave. Mani Parampettu, a social activist in Attappady, who witnessed this ritual about forty years ago, mentions that this ritual is performed only in the presence of the community members (Parampettu 2005). The gesture of preserving the skulls of the dead people symbolizes their physical presence in the hamlet. Mani Parampettu claims to have visited the sacred place years ago. Hearing about this important and unique custom of the Mudugar, I was curious to see the relics in the cave. I approached a young leader of the community, Mr. K. A. Ramu. Ramu *maash* (I call him Ramu *maash* because he calls me Rayson *maash*—meaning, master), a close friend of mine. I transcribe our conversation from memory, in which he responded "why not?" to my request to see the cave. However, he cautioned me that the path leading to the cave is a less-traveled one and hence dangerous. Determined to see the cave, I replied, "if you are there with me, I am sure I can reach the place without any fears and hurdles." Ramu *maash* casually mentioned, "people who do not belong to the community are not allowed to be in that place. There are guards commissioned by our ancestors to guard the place from outsiders." Quite dejected, I said, "I only want to see it, you could tell the guards that I am your friend," assuming that the guards were humans. Ramu *maash* said, "No, the guards are honey bees. They are in large numbers and would attack a person who smells differently."

There are two important ecocultural beliefs depicted here: the community's ancestral belief that the place should be guarded and that the honey bees guard the caves. Mudugar collect honey without disturbing or burning the honey comb. Honey is used as food and medicine. A favorite Mudugar dish is called *then cheera thavidu*; it is made of the bran of local spinach seeds and honey. A vast body of Indigenous knowledge exists amongst the Mudugar and Kurumbar regarding honeybees. Mudugar can even predict the seasons by observing the behavioral pattern of honeybees. When a group of men set out to collect honey in the forest, they sing songs praising the deeds of their ancestors. While doing so they call out the names of their ancestors and proclaim their exemplary deeds. (*The Story of Mudugar* 2008). It is in this context that we need to understand the relationship that the Mudugar share with honey collection and honeybees.

In essence, honey collection is not merely an activity for extracting honey but a traditional event with ritualistic performances attached to it. The purpose of this is far reaching. As mentioned by Raghu, a young leader of the Mudugar, the songs describe the various ways in which honey is collected; the preparation of remedies for bee stings and the prayers offered to the Mudugars' ancestors to request the bees to guard them while they are in the forest. The primary purpose of the performances is to pass on Indigenous knowledge on honey collection to the next generation. Along with the basic skills of honey extraction, the Mudugar's ethics of relating with the honey bees are also passed on in the form of songs and stories, which speak to the lived nature of ancestrality as an "unfolding, regenerative, relational, and creative process" (Poirier 2008, 79).

The Mudugar believe that their ancestors speak to the bees, and vice versa. This non-physical relationship is nonhierarchical, because bees and ancestors mutually understand each other's "language" and they live in the same physical setting. There is an interrelated web of physical and nonphysical entities in this two-fold relationship. In addition, the Mudugar people interact with bees and their ancestors simultaneously.

One might argue that the relationship between the Mudugar and the honeybees has an established physicality and mutual benefit—in the process of honey collection, the bees protect the Mudugars' ancestral remains, and the Mudugars' own ecological sensitivity helps protect the bees. However, one can also argue that the relationship is mediated through a metaphysical channel, that of the spirits of Mudugar ancestors with whom the bees can communicate. These "mediating practices and performances ... [are] very complex, flexible, and polysemic sign systems ... [D]eciphering and interpreting them ... are always creative acts (Poirier 2008, 76). In ensuring links to their ancestors, the Mudugar–honey bee relationship is ecocultural—both ecological and cultural knowledge is essential to these engagements. Thus, partaking in this sacred cultural ritual is a means by which the Mudugar preserve their ecocultural rootedness, passing on the stories of ethical responsibility required to maintain their ancestors' remains and spirits as well as healthy populations of bees.

Sacredness as Ecocultural

The last story that I will narrate further illustrates the complexity of a nature-culture-sacred nexus that is central to the Mudugar *tinai* worldview. February is a festive month for the Mudugar, as they celebrate *Sivarathri*—the festival of lights associated with the Sanskritic legendary figure, Siva. All the tribal communities of Attappady come together to organize this festival, although the festival is financed and thus dominated by the Irular. The Siva temple in Chemmannur, near the Malleswaran Hill, is the central space for the festivities. I realized, after a couple of months into my stay, that there is another side to the story on Sivarathri. Velli, a Mudugan from the Aanakkallu hamlet, mentioned that "this was a festival dedicated to Malleswaran."

Figure 10.1 Malleswaran *Mudi*—a view from the hamlet, Mele Thuduki, 2010. Source: Courtesy of Sachindev P. S.

> Sivarathri is not our festival and the Siva temple is not our temple. We have our Malleswaran temple at the foot of the mountain. The festival was originally held there. But now, though there are some rituals practiced in the Malleswaran temple, the Siva temple, across the road, is in the limelight. The Malleswaran festival is a remembrance of our origin. Thus people outside the community are not allowed to climb the Malleswaran *mudi* ["cliff"]. The custom is that every year, some of us from our community and the Kurumbar, after observing a month of abstinence from eating meat, having sex, and consuming alcohol, go to the top of the Malleswaran *mudi* and light a bonfire; viewing the fire is considered to be a blessing for the people in Attappady. We spent the entire night on the top of the cliff, talking to our ancestors, serving them food, and returning with the sacred water from the cliff, which is then distributed to everyone, irrespective of communities and religions.
>
> (Velli 2005; translation mine)

It is after hearing Velli's account that I understood the deep relationship shared between the Mudugar and the Malleswaran *mudi*, which is both a parent and deity. In her work on Latin American Indigenous cosmopolitics, de la Cadena suggests that such sacred mountains are considered "earth-beings" for Indigenous people (de la Cadena 2010, 357). Any beliefs or rituals attached to the sacred mountain "enact the respect and affect necessary to maintain the relational condition between humans and other-than-human beings (de la Cadena, 2010, 341). In the Mudugar community a similar respect is evident, as seen from this local song:

> Oh! My Mudugan son
> Oh! My Mudugan son
> My black-hill Mudugan child,
> Are you born of the fissured rock, like your ancestors?
> Oh! My Mudugan son
> My son was first laid on a bed of *chaama* leaves
> Then on a bed of *raagi* leaves.
> Oh! My Mudugan son
> My son was gifted with all kinds of food,
> With all kinds of rice.
> Oh! My Mudugan son
> He was presented with all kinds of flowers
> Oh! My Mudugan son
> Everyone stood beside him to hold his hand
> Oh! My Mudugan son
> Oh! My Mudugan son
>
> local lullaby; translation mine

This is a song sung by Ms. Sindu, age nineteen, from the hamlet of Thunduur. It is a popular lullaby familiar to all the Mudugar. This is why the Mudugar believe that the tribe is born of a fissured rock, which is an ancestral entity. The Mudugar believe that the "fissured rock" is a living entity and the mother of the community from which the tribe originated. The rock, of course, is attributed with the qualities of a human being.

Malleswaran, the black rock/hill is a deity as well as the ancestor of the Mudugar. By attributing human qualities to the hill, the Mudugar reinstate that particular rock as their ancestor. There is a *tinai* continuity between the natural, cultural, and supernatural phenomena (the Mudugar, myths, and the rock-hill) that demonstrates the rich cosmovisions of the Mudugar. Their reverence they show in worshipping is also a political act against the imposed and imported rituals of the festival of Sivarathri brought by immigrants into their community.

The Stories and *tinai* Documentation as "Deeper" Ecology

The four aforementioned stories evince elements of rootedness in the land. Maaman considering me as an outsider and Ramu *maash* saying so outright are acts of political resistance to centuries of injustice. That both do so with attention to the local ecocultural resources of their home suggests how *tinai* is cosmopolitically rooted. *Paan*, bees, and their related cultural ramifications symbolize this rootedness. The Mudugar's conflict with the Irular and other settlers in the celebration of the Malleswaran festival is the politics of belonging—implicitly an ecocultural rootedness that links to ancestral beings in their environment as allies against settlers. As shown in these stories, natural elements of the environment such as honeybees and Malleswaran *mudi* have supernatural/sacred manifestations. It could thus be concluded that the underlying principle of the Mudugar life is the strong bond between natural, cultural, and sacred elements of life—this is a distinct Indian expression of Indigenous cosmopolitics, *tinai*, where one element transforms to the other, yet retains its inherent value and identity, forming, as Selvamony calls it, part of a nexus (Selvamony 2007, xv).

Identifying/realizing this unity of the natural, cultural, and sacred is crucial to understanding the Mudugar identity and their subtle but nonetheless ever-present use of *tinai* as a way to negotiate the politics of their world—whether with supernatural beings or incoming settlers. It took me more than two years to identify this underlying principle of Mudugar life. As an ethnographic fieldworker, the oral literature that I had documented before embracing the Mudugar's philosophy of life and relationships was as a deep ecologist, Arne Naess might suggest, at "an artificial or preliminary level of communication" (Naess 1995a, 152). Such communication is, as Naess writes, seen in ecologists who have a shallow, anthropocentric attitude toward the environment and who fail to understand the deeper network or relationships that organisms have between them. Naess proposes a perspectival change, a deep ecological change, for such ecologists.

Though Naess refers to scientists as "ecological fieldworkers" (151), his concerns have relevance to this essay. In my ethnographic fieldwork, I was undergoing a process of transformational identification that in the spirit of deep ecology helped me realize ecological sensibilities of which I had been hitherto unaware. However, this transformation was different from the "identification" that Naess proposes in his essay "Self-realization: An Ecological Approach to Being in the World" (1995b), because it looked beyond the immersion of self or individual in nature to the value of a *community's* ecocultural immersion in the land. My identification was with the Mudugar people, their beliefs, social systems, myths, stories, and relationships with specific organisms and natural entities such as water and land. In essence, my identification is distinct from Naess'. Selvamony's critique of deep ecology is pertinent here. He says that the focus of deep ecology is rooted in Western science (Selvamony 2007, xiv). Naess repeatedly mentions this (Western) scientific outlook. However, what I identified with the Mudugar included the hills, waters, land, *and* myths, stories, and tribal perspectives.

As a form of identification, *tinai*-realization, and with it, *tinai* documentation, is thus the recognition and respect for the nexus of nature, culture, and sacred as expressed by communities whose ecocultural ties imbue a sustainable relationship and reverence with their environments, which is a form of political negotiation that helps frame their relations to nonhuman and human alike. While I call it ecocultural Self, faithful to Naess' nomenclature, I believe it corrects some of the problems that Naess' ideas represent. Specifically, it makes room in ecocritical discourse not just for "nature" but also for the voices and ecological knowledge systems of communities that have long been sidelined by Western hegemonies of power and scientific knowledge production (see for example, the arguments of Guha 1998, and Selvamony 2007). It alerts us to their politics of living. Most importantly, an understanding of the *tinai* ecoculture of the Mudugar has offered this documentarian insights into socioecological orders that challenge the neoliberal and capitalist agendas of India's aggressive development models, which turn a blind eye on the cultural richness of the community. As de la Cadena suggests, "'ethnic politics'" demanding 'cultural rights' may open a discussion and even articulate the need to include the Indigenous in politics" (2010, 350). Yet, unfortunately, within the context of India, as with Indigenous communities all over the world, *tinai* beliefs are often "honored only when they do not express an epistemic alternative to scientific paradigms (ecological and economic) and their cognate policies, working toward the production of the common good (productive efficiency, economic growth, even sustainable development) designed to satisfy a homogenous humanity benefiting from an also homogenous nature" (350). The self-other perspective that the NGOs, GOs, and the settlers have adopted in developmental activities and other engagements with the tribal people in Attappady are often dismissive of Indigenous worldviews as political engagement. In contrast, an informed *tinai* realization as Indigenous cosmo*politics* can help bridge this dichotomy between the tribal people and the settlers and their social, ecological, political, and cultural structures for a more sustainable and just future.

Note

1. The Irular Population, according to the Indian government's 2011 census, is eighteen times greater than the Mudugar population. As the literacy rate among the Irular has dramatically increased from 11.75% in 1981 to 29.3% in 1991 and is expected to rise up to about 50% in 2021, the Irular have taken up government jobs and assumed local political power, discriminating against the Mudugar and the Kurumbar. Surprisingly, the Irular community has managed to climb the social status ladder better than the Mudugar or Kurumbar (Feroze, 2011).

References

Alex, Rayson K. 2014. Introduction to *Culture and Media: Ecocritical Explorations*, ed. Rayson K. Alex, S. Susan Deborah, and Sachindev P. S., 1–9. Newcastle upon Tyne: Cambridge Scholars Publishing.

Ayyappan. August 2012. Personal interview by author. Video recording. Attappady,-Kerala, India. 12 August.

Bate, Bernard. 2009. *Tamil Oratory and the Dravidian Aesthetic: Democratic Practice in South India*. New York: Columbia University Press.

Chopp, Rebecca S., and Mark Lewis Taylor, eds. 1994. *Reconstructing Christian Theology*. Minneapolis, MN: Augsburg Fortress Publishers.

Damrosch, David, ed. 2013. *World Literature in Theory*. Chichester, UK: Wiley and Sons.

de la Cadena, Marisol. 2010. "Indigenous Cosmopolitics in the Andes: Conceptual Reflections beyond 'Politics.'" *Cultural Anthropology* 25 (2): 334–70.

Feroze, M. and K. P. Aravindan. 2011. "Sickle Cell Disease among Tribals of Attappady." Last modified 11 March 2011. Kerala, India: Kerala Research Programme on Local Development (KRPLLD). Accessed January 14, 2016. http://www.cds.ac.in/krpcds/report/feroz.pdf.

Franco, F. M., and D. Narasimhan. 2008. "Ecological Elements in the Songs of the Poraja and the Ancient Tamils." *Indian Journal of Ecocriticism*. 1 (1): 49–54.

Guha, Ramachandra. 1998. "Radical American Environmentalism and Wilderness Preservation: A Third World Critique." In *The Great New Wilderness Debate*, ed. J. Baird Callicott and Michael P. Nelson, 231–37. Athens: University of Georgia Press. Accessed 15 January 2016. http://isites.harvard.edu/fs/docs/icb.topic986727.files/Guha%201989.pdf.

Ingram, Annie Merrill. 2013. Preface to *Words for a Small Planet: Ecocritical Views*, ed. Nanette Norris, ix–x. Lanham, MD: Lexington Books.

Kabeer, Naila. 2012. *Inclusive Citizenship: Meanings and Expressions*. London: Zed Books.

Karat, Brinda. 2015. "A Glass Half Empty for Adivasis." *The Hindu*. May 2. http://www.thehindu.com/opinion/op-ed/a-glass-half-empty-for-adivasis/article7162260.ece/.

"Low literacy, Poverty Still Haunting Tribal Women, Say Experts." 2014. *The Hindu*. February 8. Accessed January 15, 2016. http://www.thehindu.com/news/cities/Vijayawada/low-literacy-poverty-still-haunting-tribal-women-say-experts/article5667435.ece/.

"Major Tribes in Kerala." 2016. *Focus on People*. Accessed January 14, 2016. http://www.focusonpeople.org/major_tribals_in_kerala.htm.

Naess, Arne. 1995a. "The Shallow and the Deep Long-Range Ecology Movement: A Summary." In *Deep Ecology for the 21st Century: Readings on the Philosophy and Practice of the New Environmentalism*, ed. George Sessions, 151–55. Boston, MA: Shambhala Publications.

———. 1995b. "The Deep Ecological Movement." In *Deep Ecology for the 21st Century: Readings on the Philosophy and Practice of the New Environmentalism*, ed. George Sessions, 64–84. Boston, MA: Shambhala Publications.

Parampettu, Mani. 2005. Personal interview by author. Attappady, Kerala, India. 7 January.

Poirier, Sylvie. 2008. "Reflections on Indigenous Cosmopolitics-Poetics." *Anthropologica* (1): 75–85.

Rooney, Dawn F. 2014. "Betel Chewing in South-east Asia." *Rooney Archive*. August 24. http://rooneyarchive.net/lectures/lec_betel_chewing_in_southeast_asia.htm.

Sakthivel, K. 2005. "Gurusollu." *Gurudakshina*. 1 (1): 45–51.

Selvamony, Nirmal. 1990. "An Alternative Social Order." In *Value Education Today: Explorations in Social Ethics*, ed. J. T. K. Daniel and Nirmal Selvamony, 229–50.

Madras, India: Madras Christian College and All India Association for Christian Higher Education.

———. 2007. Introduction to *Essays in Ecocriticism,* ed. Nirmal Selvamony, Nirmaldasan, and Rayson K. Alex, xi–xxxi. New Delhi, India: Sarup & Sons.

Sivathamby, K. 1974. "Early South Indian Society and Economy: The Tinai Concept," *Social Scientist* 3 (5). Accessed December 24, 2015. http://www.jstor.org/stable/3516448.

Slovic, Scott. 2010. "Reflections on Literature and Environment: An Interview with Scott Slovic. " By Diana Villanueva Romero. *Ecozon@: European Journal of Literature, Culture and Environment* 1 (2). http://www.academia.edu/1967083/Reflections_on_Literature_and_Environment_An_Interview_with_Scott_Slovic.

Story of Mudugar, The. Directed by Rayson K. Alex and Arun Bose. 2008. Chennai, India: Cockroach in Cocktail. DVD.

Sudhir, Uma. 2015. "Tribals Protest against Bauxite Mining in Andhra Pradesh." *NDTV*, November 10. Accessed January 12, 2016. http://www.ndtv.com/andhra-pradesh-news/andhra-pradesh-up-in-arms-as-government-allows-bauxite-mining-1241921/.

"Tribal Women hit Hardest by Development: Study." 2013. *The Hindu*. April 23. Accessed January 14, 2016. http://www.thehindu.com/todays-paper/tp-national/tribal-women-hit-hardest-by-development-study/article4645847.ece.

Velli. 2005. Personal interview by author. Attappady, Kerala, India. 7 January.

Velluva, Sanathanan. 1999. "Dynamics of Land-use in Recently Settled Forest Areas: A Case Study of Attappady, Kerala." Ph.D. diss., Cochin University of Science and Technology.

Vidal, John. 2009. "Are We Here Just for Your Amusement?" *The Guardian*. July 25. http://www.theguardian.com/travel/2009/jul/25/tribal-adventure-ethical-tourism-jarawa/.

Visser, Sanne (de), Élise Thébault, and Peter C. de Ruiter. 2012. "Ecosystem Engineers, Keystone Species." In *Ecological Systems: Selected Entries from the Encyclopedia of Sustainability Science and Technology,* ed. Rik Leemans, 59–68. New York: Springer.

11 The Tangibility of Maize

Indigenous Literature, Bioart, and Violence in Mexico

Abigail Pérez Aguilera

Since 1994, when the North American Free Trade Agreement (NAFTA) went into effect, the Zapatista movement has attracted considerable attention, in part because of the emergence of documents such as the Sixth Declaration of the Lacandon Jungle (2005).[1] Written by the mostly Mayan Zapatistas, this document advocates for social and environmental justice, Indigeneity, and coalition building; it imagines the construction of new, inclusive futures that raise questions about relationality amongst humans and nonhuman entities. The Declaration articulates a resistance to colonial and neoliberal systems of power that have shaped Mexico's recent history, especially around issues dealing with the country's most culturally valued food and crop, corn (maize).[2] Even before NAFTA was signed, Mexico experienced turmoil and resistance within its citizenry, in large part because of governmental policies to dispossess Indigenous peoples from lands where they typically grew corn. The government implemented massive modernization programs aimed at reforming the Mexican economy (Beaucage 1998, 4).

However, after NAFTA went into effect, the struggle for subsistence and cultural rights in Indigenous communities intensified as transnational policies exacerbated forced displacement and migration and intensified oppression of women due to the rise of gender-segmented labor markets in agriculture, manufacturing, and unpaid household and care work. The Declaration of the Lacandon Jungle, authored by the Zapatistas, is one of a growing number of movement manifestos in Latin America that resist these oppressions and seeks to regenerate Indigenous agencies and livelihoods. It connects a larger set of issues surrounding global trade, economics, and gender equity with discussions about how to achieve a sustainable future that is based on environmental justice (Blaser 2004, 26).

In this chapter, I consider forms of Indigenous resistance in Mexico to colonial histories, as well as to neoliberal development and the governance structures that undergird it, such as the North American Free Trade Agreement (NAFTA). These forms of resistance are taking artistic forms that echo and support documents such as the Sixth Declaration of the Lacandon Jungle. Specifically, I turn to the work of three contemporary Mexican artists who engage the figure of corn to critique violence against women and the violence of threats to "Indigenous knowledge systems" (IKS). These art forms,

I argue, support the resilience and survival of Indigenous communities and lifeways. The first work of art I examine is Mikeas Sánchez's bilingual work *Mojk'jäya-Mokaya* (2013). It is an interactive collection of poems, with music by Zoque musician Alejandro Burguete (Ocozocoautla, Chiapas) and illustrations by visual artist Paloma Díaz Abreu (Mexico City). In *Mojk'jäya-Mokaya*, Sánchez presents poems around Zoque traditions and cosmologies that focus specifically on the violence of lost IKS and on gender violence as two significant social problems. The second work of art I examine is an installation created by Mexico City-based artists, Minerva Hernández Trejo and Héctor Cruz, who question notions of "the human" by raising questions about what we know about the relationship of maize to humans. Hernández Trejo's and Cruz's work was presented in the art exhibit *Bioartefacts: Slowly Threshing Maize* at the Museum of Contemporary Art (MACO) in Oaxaca City in 2014; it was curated by María Antonia González Valerio. I focus on two installations that were part of this exhibit: *Zea Mays y Los Otros Sentidos* and *Zm_Maquina: Trazar la Vida* (2014), by Hernández Trejo and Cruz. The exihibit presents provocative views of environmental conflicts such as dispossession of lands, food sovereignty, and forced displacement.

My analysis of both poetry and a bioart installation takes place at the intersections of science and Indigenous cosmopolitical perspectives. My focus is on illustrating how scholars and ecocritics who focus on MesoAmerican Indigenous cosmologies in contemporary arts seek to change attitudes toward women and gender in Mexico and to support Indigenous intellectuals and artists. I explore how Sánchez, as well as Hernández Trejo and Héctor Cruz, take a critical, political, ontological approach as they represent corn as a "person" that has co-evolved with humans (Blaser 2013, 552). In taking IKS seriously, I argue, these artists confront violence against humans, and specifically women, since they also oppose the repression of human rights and environmental justice. Violence in Mexico against marginalized groups has been constant and multifaceted. As a by-product of postcolonial neoliberalism, it exacerbates forced displacement and gender violence. I will link this violence to universalist and/or anthropocentric perspectives of "the human." As Conway and Singh (2011) note:

> Western-centric forms of knowledge have silenced the colonial other through their peculiar claims to universality, their systematic rejection of their own historical-geographical particularity, their discrediting of other knowledges as unscientific. (690)

In contrast, I suggest that the cosmopolitical or pluriversal perspectives explored in my analysis of these art forms put forward the Indigenous concept that there is not one world, but many, at multiple scales, from the microcosmic to the cosmic. Such worlds situate humans in relations that blur Western binaries of nature/culture, human/nonhuman, and male/female. Thus, cosmopolitical perspectives confront anthropocentric and heteronormative systems of control. I argue that, in their evocation of these blurred boundaries as they manifest around corn—as a plant, as a cultural symbol, and as a food—Sánchez's poetry and Hernández Trejo's and Cruz's bioart are important tools to reclaim Indigenous identity, agency, and food sovereignty in the face of Mexico's push to industrialize through NAFTA-like reforms These reforms represent Western notions of "progress" by promising, for example, increased agricultural and monetary returns. Though unproven, such promises, within a postcolonial setting, often become dogma for countries in the Global South (Harding 2008, 142–44).

Gendering the "Human": Ancestral Knowledge, Indigenous Literature, and Pluriverses

To understand the work of Sánchez and Hernández Trejo and Cruz and the reasons they focus some of their work on corn, one must understand that in Mexico, Indigenous cosmologies include the symbolism of "Mother Corn," or maize. Corn is considered a creator being among many Indigenous groups. Thus, food sovereignty struggles are intrinsically related to deeper understandings of ways of living and sustaining a culture, which are taught and learned, in part, through stories about Mother Corn that have been passed down through generations. These stories contain knowledge accumulated throughout the Holocene period, or, in other words, over the course of the last 12,000 years, the same period of time during which "teosinte," an ancient grass that was domesticated into the reliable seed and food that corn has become (Adamson 2012a).

The concept of food sovereignty that has emerged in the recent Zapatista movement has been linked to teosinte, the grass that is considered the "mother of corn," and widely discussed in different disciplines, including ecocriticism and political ecology, as it relates to the Latin American (and Global South) experience (Adamson 2012a; Blaser 2013). "Food sovereignty" can be defined as the capacity of communities to decide on the use of land and labor for sustainable production of food for subsistence. In autonomous, self-reliant, noncapitalist socioecological systems, agrarian and resource politics are not dependent on private interests or on neoliberal approaches that contrast international trade to food subsistence and poverty alleviation (Altieri and Nicholls 2012, 24). Hence, the food sovereignty movement, by calling corn into the political realm, advocates modes of living that challenge over-consumption and neoliberal developments that put older lifeways and traditional methods of farming at risk.

Documents such as the Sixth Declaration of the Lacandon Jungle seek to make clear the relevance of corn and traditional food systems in Mexico and their value in revitalizing Indigenous cosmologies and IKS in movements against gender violence and for food sovereignty. Both Indigenous cosmologies and IKS focus on the relationship of humans and nonhumans, revealing this relationship to be a series of entangled interactions and dependence, oftentimes mediated by relations of production, accumulation, and race and gender relations within a defined and constrained territory. It is these relationships that Mexican artists, including Sánchez and Hernández Trejo and Cruz, seek to explore.

Maize and Gender in the Revitalization of Indigenous Literature: The Work of Mikeas Sánchez

Mikeas Sánchez is a Zoque writer and radio announcer who works for an Indigenous radio station in Chiapas.[3] Her work is shaped by the oral Indigenous traditions of the Zoque people. With a history of colonialism and displacement (like that of other ethnic groups and *pueblos* of Mexico), the Zoque are situated geographically in the states of Chiapas, Tabasco, Veracruz, and parts of Oaxaca (Báez-Jorge 1973). Before colonization, the Zoque people's territory extended to the border of what is today Guatemala. Most archeological evidence of precolonial Zoque culture dates back to 3500 bce. The Zoque call themselves *O' de püt*, which means "people of the language," "authentic," or "truthful." In 1982, the eruption of the Chichonal Volcano displaced the Zoque, which caused them to redefine their homeland, and consequently, their relationship with other Indigenous and mestizo populations. The Zoque language is among the sixty-four Indigenous languages spoken in Mexico. The history of the Zoques is closely related to that of the Mixe people, and they have been grouped into the misnamed *Populuca* ethnic group. Both groups have been oppressed in Mexico, mostly as a result of systemic racism, violence toward Indigenous cultures, and gender violence (Báez-Jorge 1973).

Sánchez's collection of poems, *Mojk'jäyä/Mokaya,* is divided into four sections: "*Ore'yomo,*" "*Mokaya,*" "*Wejpäj'kiu'y* and Naming Things," and "*Mojk'Jaya.*" Sánchez alludes to Zoque traditions and calls Zoque Indigenous cosmologies (including the figure of Corn) into her reflections and insights. *Mojk'jäyä/Mokaya* presents a series of evocations of the context of cosmological thinking of Zoque oral tradition and interconnections with Sánchez's experiences of growing up in Chapultenango, Chiapas, as well as of being an Indigenous woman in Mexico. The name of the opening section, "*Ore'yomo,*" is the word for Zoque women. In this section, the figure of *Oko'chuwe*, the "grandmother/old fearsome woman/*warm lady*" is presented as "the knower of good and bad" and "the mother of pleasure and pain." In these multifaceted representations, *Oko'chuwe* is contrasted to the visions of benevolent women figures, such as Our Lady of Guadalupe, a figure of forgiveness and passivity. In "*Ore'yomo,*" poems allude to the experiences of young women in their Indigenous communities and explore the connections to their Indigenous roots. For example, the poem "*Metza*" ("Two") of the section

"*Ore'yomo*" presents a female figure of healing who resists the elimination of Indigenous traditions:

[Ore'yomo]	[Ore'yomo]
[Metza']	[Metza']
[Dos]	[Two]
Muchacha que escondes bajo tu falda	Girl, you hid under your skirt
los secretos más exquisitos	the most exquisite secrets
Muchacha que lloras porque desconoces tu origen	Girl, you cry because you know not your origin
Diosa milenaria	millenary Goddess
muchacha	girl
ven a cantar conmigo	come to sing with me
ven a olvidar esto que nos hiere	come to forget this that hurt us
esta espinita que se nos encarna	this little thorn that is incarnated
ven hermana mía	come my sister
ven a maldecir conmigo	come to curse with me to
a todos aquellos que escupieron sobre nuestro origen	all those that spat over our origin
(Sánchez 2013, 15)	(Sánchez 2013, 15)

This female figure is recurrent throughout Sánchez's work in different forms, including one of a deeper connection with her corporeality and her Zoque identity. In spite of different symbols of motherhood and femininity across MesoAmerica (both as punisher and giver of life) (Cruz 2012), the figure of the "Girl" in this poem reflects the relevance of gender and nature within Zoque cosmopolitics, symbolizing the collective work of the community and its relation to the land. Besides her status of *mother* and deity, Mother Corn is a felt presence in many Mexican mestizo and Indigenous communities (Adamson 2012a).

Sánchez calls Mother Corn into her work as she reflects on the lives of the Zoque mestiza women and the politics of place, identity, and ethnicity by incorporating pluriverses, based on Indigenous cosmopolitics, and a critique of the human and nature divide and its relationality to the "natural world." In "*Tumä*," ("One") the first poem from the section named "*Mojk'jäyä*" ("Flower of Maize"), the author evokes the oral tradition of the Zoque people and their relation to the land and to agricultural practices. These oral traditions embody Indigenous scientific knowledges about the relation of the human to maize. In "*Tumä,*" the figure of the "flower of maize" (*Mojk'jäyä*) is presented as a complement to the male figure of maize. This shows that the Zoque understand corn to be a monoecious plant, meaning that it grows its male and female parts on the same plant. This scientific knowledge shapes approaches to "cultivating the land" and to interacting with other humans and nonhumans. While in Zoque cosmologies the corn plant represents a blurring between for all living and nonliving entities, here Sánchez also draws attention to a blurring between male and female genders. Sánchez writes:

Mokaya	**Mokaya**
Uno	One
Tüma	Tüma
Soy Mokaya	I am Mokaya
soy hombre y soy	I am man and
mujer	woman
Mojk'jäyä	Mojk'jäyä
la flor del maíz	the flower of maize
la palabra cantada	the sang word
la dolorosa palabra	the painful word
cultivo la palabra	I cultivate the word
cultivo la tierra	I cultivate the land

This poem illustrates that an essential part of Amerindian cosmologies is sexuality—a healthy part of the life cycle and reproduction—including eroticism, since sexuality is concerned not only with human reproduction but also with the reproduction cycles of nonhuman "persons," such as corn, which is a monoecious plant.[4]

In the poem "*Tumä*" ("One") from "Mokaya", the cultivation of land is in charge of *Mokaya*, a two gender entity, as a caretaker of the land. This is connected to the poem "one" or Tumä of the section "Mojk'jäyä":

Mojk'jäyä	**Mojk'jäyä**
Tumä	Tumä
Uno	One
Soy mujer	I am a woman
y celebro cada pliegue de mi cuerpo	and I celebrate every fold of my body
cada minúsculo átomo que me forma	every minuscule atom that forms me
y donde navegan mis dudas y mis esperanzas	and where my doubts and hopes sail
Todas las contradicciones son maravillosas	All the contradictions are wonderful

porque me pertenecen	because they are mine
Soy mujer y celebro cada arteria	I am a woman and celebrate every artery
donde aprisiono los secretos de mi estirpe	where I enclose the secrets of my lineage
y todas las palabras de los *ore'pät* están en mi boca	and all the words of the *ore'pät*[5] are in my mouth
y toda la sabiduría de las *ore'yomo* están en mi saliva	and all the wisdom of the *ore'yomo*[6] are in my saliva
(Sánchez 2013, 70–71)	(Sánchez 2013, 70–71)

In these two poems, "Tumä" and "Mokaya," Sánchez presents both the land and the female body as elements of nature mediated by Zoque cosmologies. Hence, that the relationship between human with nature is one of interdependence between humans and nonhumans, invisible and visible entities. Thus the female body is indeed a *political territory* (Gómez Grijalva 2012), where the women's body is not marginalized, but placed at the center. This placement confronts a genderized state-sponsored violence and an obliviousness of the eroticism, sexuality, and knowledge that the female body carries in itself. As ecocritic Stacy Alaimo notes:

> Historically, nature has been mapped by mind/body dichotomies that are coded by gender and race, associating women and people of color with abject bodily resources. Negotiating such an ideologically mined terrain is extremely difficult, especially when the idealized version of nature seems to be complicitous in maintaining its mirror image. (1996, 50)

In Poem 7, "Kuyay," in the section "Mojk'jäyä," Sánchez is concerned with denied sexual and erotic experiences for women, especially women of color,[7] for whom the exercise of sexuality is oppressed by conventions of gender and ethnicity. Sánchez presents a series of poems on female sexuality and eroticism that does not place the female body in a heteronormative state of passivity. On the contrary, the poem *Kuyay* depicts women in the exercise of their sexuality and embracing their dignified bodies. Symbolically, Sánchez is opposing subjugation of women and the forced heterosexuality required by the nation-state (Curiel 2013). Hence, the gendered bodies embedded within Sánchez's poetry are an invitation to reflect upon the fact that sexuality and gender have been shaped under a modern/colonial gender system (Lugones 2007). The reading of Sánchez's poetry is an invitation to deconstruct the imposed hegemonic sexuality, eroticism, and relations of the body.

Mojk'jäyä
Kuyay
Siete

Me nombro y hablo por todas las mujeres
que aún se duelen por su sexo

por rodas aquellas que todavía callany
aborrecen la palabra deseo
a ellas ofrezco mi espíritu
perfumado con flores de mayo
con ellas celebro mi dolor y mi gozo

(Sánchez 2013, 89)

Mojk'jäyä
Kuyay
Seven

I proclaim myself and speak for all the women
that still are hurt because of their sex
all those that still are silent

and despise the word desire
to them I offer my spirit
fragrant with May flowers
with them I celebrate my pain and my joy

(Sánchez 2013, 89)

The Indigenous woman of color's body is reframed and placed within Indigenous perspectives on the bodies of women and sexuality. MesoAmerican cultures place the human body within cycles of life and death, and include sexuality as a vital part of the cosmos. Hence, Sánchez's work relies on life cycles, the agency of nature. The Indigenous body is liberated and reaffirmed in its beauty, a worthiness embedded within Zoque cosmologies.

Mikeas Sánchez's work illustrates that ancestral and traditional knowledge, or IKS, including agroecological knowledge, has ecocultural relevance in the present. Informed by the monoecious characteristics of corn, Sánchez deconstructs the divisions of human/animal/inhuman and gendered, heteronormative understandings of female/male. Clearly, maize is both materially and culturally significant in Indigenous environmental justice movements and literary arts that are connected to Indigenous struggles for autonomy, food sovereignty, women's rights, Indigenous language preservation, and the revitalization of IKS in Latin America. Yet, the skepticism with which Indigenous IKS has been viewed within current power structures in Mexico parallels a general lack of attention and undervaluation of the intellectual work of Indigenous writers in Mexico. As Miguel Léon-Portilla (2001) writes, Indigenous cosmologies and oral traditions in MesoAmerica (as well as in the rest of the Americas) are often considered just a remembrance of the days before the colonization or a romanticization. These attitudes hinder Indigenous intellectuals from disseminating their work and bringing wider recognition to Indigenous knowledges as both scientific and cosmopolitical understandings that are valid. Léon-Portilla (2001) comments:

> It is true of course that the impact of Western civilization affected the living patterns and thinking of the original peoples of the Americas.

> But it is also true that while the imposition of foreign culture hurt the innermost cultural self of the MesoAmericans, it did not efface it. Different cultures have co-existed, influencing each other, shaping the "face and heart" (a Nahua metaphor meaning character) of what is today Mexico. (14)

Ecocritics who focus on Mesoamerican (and Latin American) poets and artists and the significance of the recuperation of Indigenous cosmologies in contemporary genres have the opportunity to change negative attitudes about Indigenous cultures that have been pervasive in Mexico and support Indigenous intellectuals and artists. In doing so, ecocritics also help reframe Western binaries of nature/human relations. According to León Portilla (2001):

> [it] would be unthinkable for the MesoAmericans, even today, to think of themselves as unattached entities, kinless, isolated in any way. They understand that the cycles, feasts, and religious ceremonies during the solar year help them immensely to reinforce the vital feeling of belonging to a sacred time and space. This ethos is the realm where humans are born, establish links of close relationship with others, work, marry, have their children, and fearlessly accept their own death as a point of encounter with Her, Our Mother/Him, Our Father. Such beliefs and attitudes are very far from those of modern Western culture. (15)

Sánchez's work illustrates that many writers in Mesoamerica are engaged in a new appreciation of IKS. Montemayor (1998) suggests that it is necessary to look at IKS as *arte de la lengua* (art of the tongue), which includes oral and written literary traditions that often fall outside Western aesthetic conventions. The links between nature, Indigeneity, and "ecological thought" in MesoAmerican literatures—and Indigenous literature from the Americas—are being integrated into a corpus of contemporary global literary arts that are incorporating Indigenous cosmologies and alternative ecological understandings to Western binaries of nature/human, object/subject.

It is these binaries and their destructive effects on people and ecosystems alike that artists such as Sánchez, ecocritics, anthropologists, political ecologists, scientists, environmental activists, and ethnologists resist by bringing forward perspectives of pluriverses that recognize the multiplicity of "views," which also include nonhuman actors in the realm of politics (de la Cadena 2010; Carrillo Trueba 2006). Sánchez encompasses both humans and nonhumans and her work is dedicated to *Nasakobajk*, which is Mother Earth in Zoque cosmology and the protective spirit of the world. The name *Nasakobaj* comes from *Nas* (meaning "head") and *kobajk* (meaning "land"); therefore, *Nasakobajk* is the head of the land (Lisbona 2004, 226). In Sánchez's work there is no separation between the *natural* as an isolated, or wilderness space and the space of the *human*. For example, in Poem 6

("Tujtay"), Sánchez writes of "the prayer of the peasant," in which a man asks not for favors but to understand the temporality of the human body and the limitations of human understandings of the nonhuman.

Mokaya
Seis

La oración del sembrador reza
"Kujkiki'
esta es la palabra que me enseñaron
los que caminaron la tierra antes de mí
Kujkiki'
será la palabra que dejaré a mis hijos
el día de mi muerte
Porque no excederé mi paso por la tierra
ni abusaré del placer ni del dolor

Con una mano brindo mi ternura

con la otra recibo el calor del sol"

(Sánchez 2013, 36–37)

Mokaya
Six

The prayer of the sower goes
"Kujkiki'
This is the word that I was taught
The ones that walked the land before me
Kujkiki'
It will be the word that I will leave to my children
The day of my death
Because I will not exceed my passage on this earth
Nor abuse of the pleasure and pain
With one hand I offer my tenderness
With the other I receive the warmth of the sun"

(Sánchez 2013, 36–37)

In this prayer, there is a continuity between humans and earth-beings, or the "ones that walked before," even in contemporary times. Cosmic entities such as the sun do not represent a life-death binary but are seen to exceed a time-space continuum that falls outside of Western (and Abrahamic) conceptions of an "end of time."

Sánchez's poems illustrate how Mexican artists are engaged in forms of critique that support and reflect the goals and aims of documents such as the Sixth Declaration of the Lacandon Jungle. Both political document and poetry oppose violence against women that takes the form of heterosexuality enforced by the nation-state or threats to IKS that resist binary separation of nature and culture.

Zea Mays y los Otros Sentidos and *Zm Maquina*

Opposition in Mexico to the loss of IKS, which is another form of violence against Indigenous peoples, is also taking the form of art installations. Since major debates about transgenic corn in Mexico are unfolding mostly in Indigenous territory, opposition to transgenic corn is a struggle that is entangled in debates about food sovereignty, Indigenous understandings of Mother Corn, and Indigenous rights. As conceived by many Indigenous communities, this is a multifaceted conflict that involves human and "non-human substances, systems and beings" (Latta 2014, 325) where lives are at stake and unconventional agents, such a sentient plant entities and sun and earth-beings, are incorporated into the realm of politics. In response to these debates and with reference to such sentient entities, two of the installations in the exhibit, *Zea Mays y Los Otros Sentidos* and *Zm. Maquina: Trazar la Vida* (2014), by Minerva Hernández Trejo and Héctor Cruz, confront what Bruno Latour has called a "modern constitution," meaning a "regime of life that create[s] a single natural order and separate[s] it from the social by creating an ontological distinction between things and humans that it purports [as] universal" (Latour, quoted in de la Cadena 2010: 342). Both installations are part of the MediaLab 2014 artist collective at Centro Multimedia-CENART. Each of these works raises questions about the ways humans interact with plants by mediating and interpreting this relationship with a machine created by Hernández Trejo. The purpose of both installations, like other Indigenous works of art described by Adamson (2012b, 44), is to "re-vision the planet as a cosmos of multispecies communities existing in intimate, entangled relations." The implications of this re-visioning and uncovering of different pluriverses, entail profound changes, as they discuss and seriously consider alternative paradigms to Mexico's modern reforms intended to promote "progress," which are rife with eco-social injustices (see, e.g., Beaucage 1998).

The work of Minerva Hernández Trejo and Héctor Cruz, like that of Sánchez, is part of an initiative by contemporary artists in Mexico to explore the interaction of humans with "persons," or the more-than-human world. This exploration can be seen in art installations, photography, digital arts, and corporeality studies, among other arts. In the collective exhibit *Bioartefactos: Desgranando Lentamente un Maiz* (2014) ("Bioartefacts: Slowly Threshing Maize"), shown from June to September of 2014 in the State of Oaxaca, Hernández Trejo and Cruz sought to engage concepts of "living matter" within Indigenous cosmologies. Thus, they incorporate MesoAmerican storytelling and writing traditions about different entities that comprehend and promote understanding of the cosmos (Reyes 2008). To express these concepts, both artists focus on corn, with its associated cosmologies, biology, and politics.

As suggested in my analysis of Sánchez's poems, in Mexico maize is both a tangible and intangible presence, with known cultural and political meanings within environmental justice struggles, including struggles for human rights and food sovereignty. There are also tangible politics connected to corn. Before the revision of the Mexican Constitution, Article 27 guaranteed all citizens a "tortilla a day," thus mitigating food insecurity for the poorest and most vulnerable. Since the passage of NAFTA and the revision of Article 27, citizens of Mexico are no longer guaranteed this "right" (Adamson 2002, 49–50). Meanwhile, NAFTA gives multinational corporations the legal right to out-compete Indigenous farmers in international markets, leading to dramatically increased food insecurities and displacement of Indigenous people from their communal farms (Adamson 2012a). Thus, discussions about food sovereignty open the door to both national and civil debates and to Indigenous cosmopolitical conversations about interactions between and among humans and nonhumans. As Beatriz Cariño (2010), an environmental activist, has observed:

> Our fields now are the scenes of ruin and disaster, victims of indiscriminate commercial opening, genetically modified crops, the ambitions of the multinationals; this has consequently caused the forced migration of millions of our brothers and sisters who, in the words of my grandfather, "have to leave in order to remain." (34)

Linking economics to human and ecological struggles for autonomy and rights, the "political ecology of maize," as Kathleen McAfee terms it (2008, 149), reflects a complex set of connections that are made visible in the cultural, biological and political debates emerging around transgenic crops. Environmental justice movements connect transgenics to issues of biosafety, food sovereignty, and the "coloniality of nature" (Escobar 2008, 61–92). The concept of coloniality of nature has been developed by Arturo Escobar (2008) and expanded by Anibal Quijano's concept of *coloniality of power* (2000). These notions contest the modern capitalist system, its impacts on labor relations, its structures of gender and racial categories, as well as its systems of epistemic coloniality (or a colonialiy of knowledge). When maize (along with other nonhuman agents) becomes a commodity under the modern system and its importance is assigned a market and profitability value, the result is a degraded, exploited, and fragile local food system dependent on the external and global powers. Often the problem that arises is that communities lose self-sufficiency.

In such implicitly violent systems, maize, as a food and as a culturally meaningful plant, is reduced by the dominant Western society to a commodity. IKS that once enabled alternative perspectives on ethical agroecological farming practices is threatened and may be lost, given the displacement of Indigenous farmers. The relationship between corn farmers, or *campesinos*, and their agroecological practices and the cosmovisions these practices entail concerning diverse types of sentient beings, such as Mother Corn (and nonsentient beings, such as the soil and how to care for it), are threatened. With agricultural modernization, the active everyday of grassroots "translation" or "articulation" of IKS is now replaced by hierarchical top-down expert knowledge and monocultural corporate farming. Humans and more-than-humans become defined through Western scientific knowledge, while specific Indigenous knowledges are lost. As one example, we can consider *Ustilago maydis*, or "corn smut," called *huitlacoche* in Nahuatl, in its contexts in Mexican IKS. *Huitlacoche*, a kind of fungus, like mushrooms, was associated with a pre-conquest diet and was part of an ancient food system and food culture for people in Mexico (Torrero and Hernández Urbiola 2010). As other food practices were brought in by settler colonials, a diet based on Indigenous knowledge that provided food sovereignty, health, and well-being was discarded and traditions around *Huitlacoche* were lost. Acoma poet Simon Ortiz has helped to explain exactly what was lost when traditions involving corn, such as *Huitlacoche*, disappeared. In the Americas, first peoples considered their relationship to corn sacred because "the earth regenerates the human body when people eat corn and, when they die, humans return to the earth and the cycle continues" (quoted in Adamson 2012a, 228). Thus, the human body and corn, or even fungal growths attached to corn, were part of sacred human/nonhuman interdependencies.

In a similar fashion, in colonial Mexico amaranth, once the source of food sovereignty and wealth among the Aztecs, was prohibited by the Spanish Catholic settlers because of its relation with the Aztec god Huitzilopochtli, a "person," or sentient entity, associated with the sun and war. The prohibition was based on the presence and use of *tzoalli,* a paste made of amaranth, which had a sacred value. *Tzoalli* was used to make cakes in the shape of "persons" such as Huitzilopochtli for Aztecan religious ceremonies The *tzoalli* was perceived by Spanish settlers and religious authorities as too similar to the Eucharist, and so fields of amaranth were burned and banned (Adamson 2012a; Velasco Lozano 2001, 60). Once the basis of food sovereignty, traditions surrounding amaranth and the IKS associated with this plant and food, though never lost completely, were put at risk. These two examples, corn smut and amaranth, reveal the complexity involved in study of the relationship of nonhumans and humans in the context of two conflicting worldviews, one based on the modernization of food-production processes and the other based on the connection of humans and nonhumans within living, interwoven, cultural and ecological systems (Lenkersdorf 1998).

In the work of Minerva Hernández Trejo and Héctor Cruz, maize is considered a biological artefact created by humans for human consumption and related to traditional agricultural systems, or *la milpa*, meaning "garden" (Richard 2012). In their art installations, Hernández Trejo and Cruz take up the debates surrounding transgenic corn, in which there is a manipulation of "life." Many farmers and activists, and many Indigenous groups, are concerned that changes to genes may be a transgression of life and culture and produce the "unnatural." Hence, the interest among Mexican artists in discussing *bioartefacts* in the context of Latin America and the connection of transgenic corn to the ethics of science and technology and considerations of cultural and social problems. The installation of Hernández Trejo's *Zm Maquina* was presented previously as *Desmodium Maquina* (2012) as part of the exhibit, *Sin Origen Sin Semilla* 2012–2013 ("Without Origin,

Without Seed"), an exhibit that precedes *Bioartefactos* and *Slowly Treshing the Corn* (2014). In the case of *Desmodium Maquina (2012)* research was conducted

> for the cultivation of *Desmodium gyrans*—unusual species of corn in Mexico, as well as research and experimentation on sensing photosynthesis processes in plants, the design of hybrid devices, experimentation with electromagnetic mechanisms *for data visualization and operation of philosophical theories dealing with the problems that open in the assemblages between nature and technology.*
>
> (Monreal 2012; my emphasis)[8]

In *Zm_Maquina* (2014), Hernández Trejo and Cruz have created a machine that monitors these assemblages, or more specifically, the processes responsible for the "breathing" of maize. The machine transmits data through a mechanical arm that records its results on a copper disc (see Figure 11.1, Zm_Maquina (2014). As a grass, corn takes in sunlight and carbon, and "exhales" oxygen and water, thus it "breathes." *Zm_Maquina*'s captures "data" on the breathing of corn and sets this information in to the context of controversies surrounding biotechnology and transgenic maize in Mexico. These controversies have been at the center of recent debates being conducted by locally based communities, in alliance with local and international scientists, and environmental activists. Such alliances "are often less about traditional politics and more about opening a political-economic-cultural space for multiple alternative developments" (Adamson 2012a, 249; see also Alvarez-Buylla and Piñeyro-Nelson 2013). Thus *Zm Maquina* takes its place among a large corpus of artistic production including literary production, digital humanities, performance studies in Mexico that are questioning transnational forces impacting local food systems.

By recording the biological processes of maize (*Zea Mays*) and translating its "breath" into readable data that argues for the "sentience" or "life" of the plant and its agential qualities, *Zm_Maquina* illustrates how bioartefacts can "make the living visible" (Monreal 2012) and serve as a testimony of relationship between humans and nonhumans. The machine in this exhibit, then, is collecting data on both the sentience and agency of corn.

In another related installation piece by Medialab 2014, *Zea Mays: Trazar la Vida* (2014), Hector Cruz explores how nonhumans—in this case, living *Zea Mays*, or living corn—react and interact with human presence. As Mitchell (2012) notes, "bioart almost invariably encourages this embodied engagement with the work of art in order to produce a sense of "becoming–medium" on the part of a gallery-goer—that is, a sense of how one's own body can become a part of the milieu for other forms of life" (92). In this installation, the machine is a medium between humans and earth-beings. It seeks to record, in data, a common view found in MesoAmerican Indigenous cosmologies. The machine literally records the interactions between

Figure 11.1 ZM MAQUINA (2014). Source: Courtesy of Minerva Hernández Trejo.

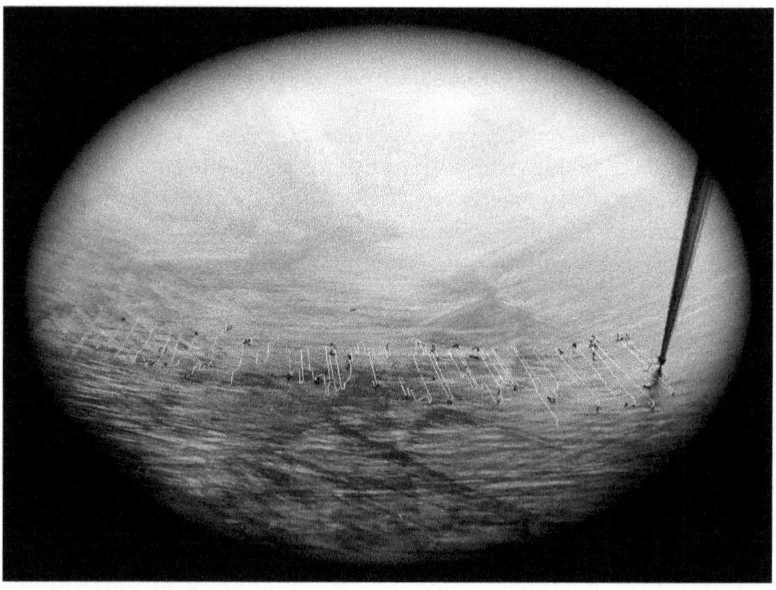

Figure 11.2 ZEA MAYS (2014). Source: Courtesy of Minerva Hernández Trejo.

humans and the more-than-human world. Both installations, *Zm Maquina* and *Zea Mays*, invite deeper discussions on the alternative ontologies and pluriversal politics of humans and nonhumans. Through this work, Hernández Trejo and Cruz (and the artist collectives BioexmachinA and Arte+Ciencia), as well as the *Colectivo Desmodium Máquina* (2014), seek to record different ontological conceptualizations: on one side, the relation of humans, plants, and artificial systems through the lenses of bioart, and on the other, a concern for the revitalization of IKS, including Indigenous cosmologies surrounding corn.

Future Possibilities

In this essay I considered how in Mexico both contemporary Indigenous writers such as Sánchez and non-Indigenous artists like Hernández Trejo and Cruz speak and write about agential conceptualizations of human and nonhuman relations. I explored the intangible political and cultural relevance of *corn* as a nonhuman agent with agential qualities within socio-environmental contexts and conflicts. The work of Sánchez and Hernández Trejo and Cruz illustrate how Indigenous and non-Indigenous cultural and artistic productions are being employed to critique systemic violence against Indigenous women, IKS, and nonhuman forms of life. In their artistic reflections on the role of corn and other nonhuman agents in food sovereignty movements (as well as environmental movements), Sánchez and Hernández Trejo and Cruz present alternative perspectives to the anthropocentric, universalist approach of how to know the "world;" they also trouble the boundaries between conventional ideas of what is perceptible and imperceptible, tangible and intangible.

The poetry of Mikeas Sánchez focuses on elements within Indigenous cosmologies, such as Mother Corn, or corn, to present and embrace not a universe, but pluriverses. This is art as a form of resistance to the loss of IKS. The bioart of Minerva Hernandez Trejo and Hector Cruz (and the projects of BioexmachinA and project Arte+Ciencia) illustrate the interaction between *Zea Mays* and human presence. All these artists mediate different kinds of knowledge where the intangible aspects of corn are expressed in terms that become "readable" to humans. By providing insights into pluriverses that have been overlooked and taken for granted or devalued, it is possible to delineate a critique of anthropocentric perspectives on life cycles, rationality, and relationality between humans and the more-than-human world.

Ecocritics, artists, and activists in Latin America focusing on or creating works of art such as these are inviting readers from both the Global South and the Global North to think about the diverse actors involved in environmental justice movements around the world. Different socioecological movements (which also embrace a critique of a gendered and racialized violence) question the legacy of the modern/colonial capitalist/patriarchal world-system (Grosfoguel 2007). As alternatives, they propose a pluriverse of diverse epistemologies and political ontologies. This movement, and the arts associated with it, show that human rights and food sovereignty movements in the Americas consider how to embrace IKS and different ontologies in a manner that materializes human perception of nonhuman phenomena and of human existence itself. It offers spaces of resistance (and contestation) as well as possibilities of denied and overlooked existences.

Notes

1. For a brief history and context of the Zapatista movement in Mexico, see Adamson 2001, 128–61; 2012a.
2. Throughout this chapter I have used the terms "maize" and "corn" interchangeably.
3. For information on XECOPA, *Voz de los Vientos* ("Voice of the Winds"), a radio station primarily for the Zoque- and Tsotsil-speaking people, see http://ecos.cdi.gob.mx/xecopa.html.
4. It is important to note that Indigenous traditions in relation to sex and gender are not constrained by a duality of women/men, female/male, or by heterosexual relationships.
5. A Zoque man.
6. A Zoque woman.
7. The concepts of "people of color" and "women of color" have not been discussed in Latin American up until now, although there is an emerging discussion of these subjects in Latin American and Caribbean feminist scholarship. However, for the purposes of this chapter and in recognition of the potentiality that transnational work on Indigenous feminisms and its convergent scholarship might achieve, I therefore use the terms. See Bidaseca (2011); Lugones (2008); and Segato (2007).
8. Both *Zm Maquina* and *Desmodium Maquina* are based on the work of scientist Sir Jagadis Chandra Bose and his machine, which is named *Desmodium gyrans* (Hernández Trejo 2012).

References

Adamson, Joni. 2001. *American Indian Literature, Environmental Justice, and Ecocriticism: The Middle Place*. Tucson: University of Arizona Press.

———. 2002. "Throwing Rocks at the Sun: An Interview with Teresa Leal." In *The Environmental Justice Reader: Politics, Poetics, and Pedagogy*, ed. Joni Adamson, Mei Mei Evans, and Rachel Stein, 44–57. Tucson: University of Arizona Press.

———. 2012a. "Seeking the Corn Mother: Transnational Indigenous Organizing and Food Sovereignty in Native North American Literature." In *Indigenous Rights in the Age of the UN declaration,* ed. Elvira Pulitano. Cambridge, UK: Cambridge University Press.

———. 2012b. "Whale as Cosmos: Multispecies Ethnography and Contemporary Indigenous Cosmopolitics." *Revista Canaria de Estudios Ingleses* (64): 29–48.

Alaimo, Stacy. 1996. "Displacing Darwin and Descartes. The Bodily Transgressions of Fielding Burke, Octavia Butler, and Linda Hogan." *Interdisciplinary Studies in Literature and Environment* 3 (1): 47–66.

Altieri, Miguel A., and C. I. Nicholls. 2012. "Agroecology Scaling Up for Food Sovereignty and Resiliency." In *Sustainable Agriculture Reviews* 11: 1–29.

Alvarez-Buylla, Elena, and Alma Piñeyro-Nelson, eds. 2013. *El Maíz en Peligro ante los transgénicos: Un análasis sobre el caso de México*. México: *UNAM-UCCS*.

Báez-Jorge, Félix. 1973. Los Zoque-Popolucas: Estructura Social. Issue 18. Instituto Nacional Indigenista, Secretaría de Eduación Pública.

Beaucage, Pierre. 1998. "The Third Wave of Modernization: Liberalism, Salinismo, and Indigenous Peasants in Mexico." In *The Third Wave of Modernization in Latin America: Cultural Perspectives on Neoliberalism* ed. Lynn Phillips. Wilmington, DE: SR Books.

Bidaseca, Karina. 2011. "'Mujeres blancas buscando salvar a mujeres color café': Desigualdad, colonialismo jurídico y feminismo postcolonial." *Andamios: Revisita de Inverstigación Social,* 8 (17): 61–89.

Bioartefactos: Desgranando Lentamente un Maíz. 2014. [Exhbitition.] Oaxaca City, Oaxaca: Museo de Arte Contemporáneo de Oaxaca. June–September.

Blaser, Mario. 2004. "Life Projects: Indigenous Peoples' Agency and Development." In *In the way of development: Indigenous peoples, life projects, and globalization*. ed. Mario Blaser, Harvey A. Feit, and Glenn McRae. London: Zed Books.

———. 2013. "Ontological conflicts and the stories of peoples in spite of Europe." Current Anthropology 54 (5): 547–68.

Cariño, Beatriz. 2010, "Book of Testimonies. The Fifth Dublin Platform for Human Rights Defenders". Accessed May 1 2016, https://www.frontlinedefenders.org/files/Dublin%20Platform%202010%20Book%20of%20Testimonies%20web%20version_0.pdf.

Carrillo Trueba, César. 2006. *Pluriverso: un ensayo sobre el conocimiento indígena contemporáneo*. Collección la Purlalidad cultural en México, Num. 11. Edited by José del Val. México, DF: National Autonomous University of Mexico (UNAM).

———. 2011. "Esquemas para Pensar." *Ciencias* 101: 24–25.

Conway, Janet, and Jakeet Singh. 2011. "Radical Democracy in Global Perspective: Notes from the Pluriverse." *Third World Quarterly* 32 (4): 689–706.

Curiel, Ochy. 2004. "Identidades Esencialistas o Construcción de Identidades Políticas: El Dilema de las Feministas Negras." *Revista Electrónica* 36.

———. 2013. *La nación heterosexual: Análisis del discurso jurídico y el régimen heterosexual desde la antropología de la dominación*. Bogotá, Columbia: Brecha Lésbica.

de la Cadena, Marisol. 2010. "Indigenous Cosmopolitics in the Andes: Conceptual Reflections beyond 'Politics.'" *Cultural Anthropology* 25 (2): 334–70.

Escobar, Arturo. 2008. Territories of Difference: Place, Movements, Life, Redes. Durham, NC: Duke University Press.

Gómez Grijalva, Dorotea A. 2012. *Mi Cuerpo Es Un Territorio Político*. In Collección Voces Decoloniales, Cuaderno 1. Bogotá, Columbia: Brecha Lésbica. https://brechalesbica.files.wordpress.com/2010/11/mi-cuerpo-es-un-territorio-polc3adtico77777-dorotea-gc3b3mez-grijalva.pdf.

Grosfoguel, Ramón. 2007. "The Epistemic Decolonial Turn: Beyond Political-Economy Paradigms 1." *Cultural studies* 21 (2–3): 211–23.

Harding, Sandra. 2008. *Sciences from Below: Feminisms, Postcolonialities, and Modernities*. Durham, NC: Duke University Press.

Hernández Trejo, Minerva. March 24, 2014 (6:02 p.m.). Comment on "Desmodium Máquina." *Minerva.org* (blog). Accessed April 20, 2016. http://minerva.org.mx/desmodium-maquina/.

Hernández Trejo, Minerva, and Héctor Cruz. 2014. *Zea Mays y los Otros Sentidos.* Installation Art. Oxaca, Mexico: Museo de Arte Contemporáneo de Oaxaca.
Latta, Alex. 2014. Matter, Politics and the Sacred: Insurgent Ecologies of Citizenship. *Cultural Geographies* 21 (3): 323–41.
Léon-Portilla, Miguel. 2001. *In the Language of Kings: An Anthology of Mesoamerican Literature—Pre-Columbian to the Present.* W. W. Norton.
Lenkersdorf, Carlos. 1998. *Cosmovisiones: Collección Conceptos.* Mexico, DF: National Autonomous University of Mexico.
Lisbona, Miguel. 2004. *Sacrificio y castigo entre los zoques de Chiapas: Cargos, intercambios y enredos étnicos en Tapilula.* Mexico, DF: Programa de Investigaciones Multidisciplinarias sobre Mesoamérica en el Sureste, National Autonomous University of Mexico.
Lugones, María. 2007. "Heterosexualism and the Colonial/Modern Gender System." *Hypatia* 22 (1): 186–219.
———. 2008. "Coloniality and Gender." *Tabula rasa* 9: 73–102.
McAfee, Kathleen. 2008. "Beyond Techno-Science: Transgenic Maize in the Fight over Mexico's Future." *Geoforum* 39 (1): 148–60.
MediaLab. 2012. *Desmodium-Máquina,* Installation art. Museo Universitario de Artes y Ciencias. MUCA Roma.
———. 2014. *Zm Maquina.* Installation art. Oxaca, Mexico: Museo de Arte Contemporáneo de Oaxaca.
Mitchell, Robert. 2012. "Simondon, Bioart, and the Milieu of Biotechnology." *Inflexions* 5: 69–111.
Monreal, Fernando. 2012. "Desmodium-Máquina." Arte+Ciencia. Accessed February 10, 2015. http://www.artemasciencia.com/exposinorigen.
Montemayor, Carlos. 1998. *Arte y Trama en el Cuento Indígena.*Mexico, D.F: Fondo de Cultura Económica.
Quijano, Aníbal. 2000. Coloniality of Power and Eurocentrism in Latin America. *International Sociology* 15 (2): 215–32.
Reyes, Luis Alberto. 2008. *El pensamiento indígena en América: Los Antiguos Andinos, Mayas y Nahuas.* Buenos Aires, Argentina: Editorial Biblos.
Richard, Analiese. 2012. "Sin Maíz No Hay País." In *Environment and Citizenship in Latin America: Natures, Subjects and Struggles* ed. Alex Latta and Hannah Wittman. New York: Berghahn Books.
Sánchez, Mikeas. 2013. *Mojk'jäya-Mokaya.* Mexico, DF: Pluralia Ediciones.
Segato, Rita Laura. 2007. *La nación y sus otros: Raza, etnicidad y diversidad religiosa en tiempos de políticas de la identidad.* Buenos Aires, Argentina: Prometeo Libros.
Sixth Declaration of the Lacandon Jungle. 2005. Schools for Chiapas. Accessed April 12, 2016. http://www.schoolsforchiapas.org/library/sixth-declaration-lacandona-jungle/.
Torrero, Esther Pérez, and Margarita Isabel Hernández Urbiola. 2010. "La Alimentación en el México Prehispánico y Actual: su Influencia en la Condición Nutricional." *La Ciencia, el Desarrollo Tecnológico y la Innovación en Querétaro. Historia, Realidad y Proyecciones*: 219.
Velasco Lozano, Ana María Luisa. 2001. "Los Cuerpos Divinos: La Utilización del Amaranto en el Ritual Mexica." In *Animales y plantas en la Cosmovisión Mesoamericana*, ed. Yólotl González Torres, 39–64. México, D.F.: Plaza y Valdés Editores: CONACULTA-INAH: Sociedad Mexicana para el Estudio de las Religiones.

12 Why Bears, Yakumama (Mother of All Water Beings), and Other Transformational Beings Are (Still) Good to Think

Joni Adamson and Juan Carlos Galeano, with illustrations by Solmi Angarita

Transformational "Persons," Ecocriticism, and Participant Observers

At the 2011 Association for the Study of Literature and Environment (ASLE) annual conference held in Lawrence, Kansas, Juan Carlos Galeano gave a plenary reading of poems from his chapbook, *Amazonia* (2012). Many of these poems playfully represent tales about "transformational beings" he heard while growing up in Colombia, near the Caquetá River, in the Amazon basin region. While his father and other community members were building houses, bridges, and furniture from the trees being felled in the rainforest, he was continually running away to hang "out in canoes with Indigenous kids and fishermen" who would talk about trees, animals, and fish "in such a sentient realm" (Fernandez 2015, n.p.). In interviews about those experiences, Galeano has said he could feel the "spirits in the air, in the trees, in the water, everywhere" (Fernandez 2015, n.p.). For over a decade now, he has been traveling back to the rivers and forests of the Amazon basin where he grew up. As a professor of Spanish at Florida State University, he oversees a student language immersion and service-learning program. In addition to overseeing this program, he continues to listen and record "oral narratives from fishermen, hunters, loggers and small town dwellers" (Fernandez 2015, n.p.). He has collected many of these stories, together with his own explanatory notes, in *Folktales of the Amazon* (2009).

As Galeano emphasized in a subsequent session at the ASLE conference, it is still common to hear local Amazonian peoples tell tales about the Mother of the Forest (Sachamama) and other "more-than-human beings" today (Galeano 2009, 11). They are told for entertainment, but also to theorize contemporary developments around globalization and accelerating environmental changes and widening social disparities. In their co-produced and co-directed documentary film, *The Trees Have a Mother: Amazonian Cosmologies, Folktales, and Mystery* (2008), Valliere Richard Auzenne and Galeano record local people in the Amazon Nanay and Itaya river region around Iquitos, Peru, telling stories about transformational beings that take the form of anacondas, trees, dolphins, birds, or clouds, or that sometimes transform themselves into the shape of humans. The locals refer to the Amazon River as *Yakumama*, which means "Mother of All Water Beings," and talk about her appearing in human form or taking the form of a giant anaconda (Galeano 2009, 13). As anthropologist Julie Cruikshank (2005, 57) emphasizes, throughout the Americas, from the Arctic to the Antarctic, Indigenous tales about significant glacial fields, rivers, mountains, or forests are gathered in oral story cycles, or "oral archives." Each Indigenous or First People or Nation group tells stories about significant geographical, animal, or plant denizens of their region in many versions. These stories have been passed down orally for generations and are still told today as "timeless narratives in a timely way" not as "myth" or as "cultural artifact extraneous to history" (Cruikshank 2005, 57).

Following Galeano's plenary session at the ASLE conference, ecocritic Joni Adamson met with him for the first time. They recognized they had many aesthetic, ethnographic, and activist interests in common. Adamson had studied with some of the most honored Native North American contemporary writers and artists, including N. Scott Momaday (Kiowa), Joy Harjo (Creek/Muscogee), Simon Ortiz (Acoma Pueblo), and Leslie Marmon Silko (Laguna Pueblo) during her years earning a doctoral degree at the University of Arizona.[1] There she had also been trained in the "participant observer" methodologies of Franz Boas, Ruth Benedict, Victor Turner, James Clifford, and Barbara A. Babcock whose metholologies are collectively known as ethnography.[2] For many years, she also taught classes at the University of Arizona for Native North American students, who taught her to understand their oral traditions and read contemporary Indigenous literatures in the context of their own experiences growing up on reservations or in cities. Adamson's experiences as a "participant observer" working in the classroom with Indigenous students and traveling with her students to their homes and communities in Arizona would have profound implications for the way that she would read and interpret Indigenous literatures and develop theories about environmental justice in the emerging field of ecocriticism.[3]

In one of her first published articles, "Why Bears Are Good to Think" (1992), Adamson began to tease out why "more-than-human" or "transformational" characters from Indigenous oral traditions should be taken seriously in the emerging field of ecocriticism. She argued that these characters—hero-twin monster slayers, hunters, bears, coyotes, snakes, or even frogs—illustrate a worldview in which "no sharp lines can be drawn dividing living beings" (Adamson Clarke 1992, 31). She explained that not all other-than-humans depicted in either the oral tradition or in contemporary novels or poetry suggest themselves as transformational. However, some species are seen to embody special abilities that suggest they move freely between more than one world (31–32). For example, a bear's ability to hibernate, a snake's ability to shed a skin, or a frog's ability to thrive both above and below the water suggest these species may be boundary-crossers and may have the ability to slip into and out of human form. They might behave like humans or they might become animal. Given these special abilities, their activities are thought to take place in "spatiotemporal frameworks" and dimensions considered "cosmic, rather than mundane" (Adamson Clarke 1992, 45n). As symbolic anthropologists and cultural theorists have long established, "transformational entities" are "good to think" because they represent the "peculiar unity of the liminal: that which is neither this nor that and yet both" (Turner 1967, 99). They are singled out, "not because they are 'good to eat' or 'good to prohibit' but because, as Claude Lévi-Strauss famously puts it (1982), they are 'good to think'" (see also Adamson Clarke 1992, 32–35; Babcock 1975, 167).

Later, in her monograph *American Indian Literatures, Environmental Justice and Ecocriticism: The Middle Place* (2001), Adamson built on her training with Native North American writers and among her students to explore why characters from Indigenous oral traditions are so frequently written into contemporary novels and poetry by contemporary Indigenous writers, poets, and filmmakers. As one example, she examined Louise Erdrich's novel *Tracks*, in which a young Annishinabe female character survives historically documented epidemics and genocide in the early part of the twentieth century in the United States. Fleur is rumored to be a "bearwalker," or transformational "more-than-human," which suggests that she may possess the strength and power of a bear, the intelligence of a wolf, and the ability to dive under water for long periods of time like the "water monster" said to live in the nearby lake (Adamson 2001, 101–12). Erdrich gives Fleur the qualities of multiple beings from Anishnaabe story archives and thus illustrates how many Indigenous artists and writers set their characters into a time-space continuum of past, present, and future. Living in the present, dealing with the oppressions and disparities enforced on Indigenous peoples by colonial legacies, the character is simultaneously a comment on the "relic" of proud ancient cultures represented by stories about bears and bearwalkers, and, at the same time, a comment on the ways modern Indigenous peoples survive and intelligently respond to conditions in a rapidly changing world. In short, Fleur acts as a "thought experiment," allowing readers and Native North American students to not only glimpse their own histories and cultures as they were in the past but to also see how individuals and communities today are bringing their stories and histories forward to support their survival, their growing strength as communities, and their planning and implementation of politics, legislation, and planning that is building their futures.

Galeano's own work with Florida State University students and his deep, abiding interest in Amazonian culture has led him to train himself as a 'neo-Boasian' (Uzendoski 2009, *xi*). While living and working among the people whose stories were finding their way into his poetry, ethnographic work, and filmmaking, he collected (and is still collecting) stories about anacondas, dolphins, and manatees said to sometimes transform themselves into humans (Galeano 2009). As anthropologist Marisol de la Cadena has explained, the names of some of these beings—Pachamama (Mother Earth), Yakumama (Mother of All Water Beings), or Sachamama (Mother of the Forest)—evoke not "mothers" but androgenous transformational "persons" that represent significant biospheric, riverine, or forest ecosystems. Their names are more correctly interpreted to mean "Source of Life" (de la Cadena 2010, 335, 350). These "persons" were made the subject of a conference organized in 2015 by Brazilian anthropologist Eduardo Viveiros de Castro called *Os Mil Nomes de Gaia* / The Thousand Names of Gaia.[4] As Donna Haraway (2015) argued in her presentation at this conference, "persons" are the names given to complex nature-culture systems, or "collected things," which give presence to an entangled myriad of temporalities and spatialities, or "entitites-in-assemblages," which, according to Haraway, include "more-than-human, other-than-human, and inhuman" (Haraway 2015, 160). They convey, according to Vivieros de Castro, a philosophy of "perspectival multinaturalism," expressed in stories about transformational beings that suggest that "the world is inhabited by different sorts of subjects or persons which apprehend reality from distinct points of view" (Viveiros de Castro 1998, 469). These philosophies speak of humans, animals, plants, and spirits participating in the same world, although with different sensory apparatus constituting not just multicultural (human) worlds that imply a unity of nature and a multiplicity of cultures but also multinatural worlds that imply corporeal diversity and its attendant diversity of perspectives (Viveiros de Castro 2004, 467).

Donna Haraway takes Lévi-Strass's formulation even further to say that transformational beings are good to think not only because of their liminal characteristics but because they are "good to live with" and they remind us that there would be no "human" without the other-than-human (Haraway 2015, 160). Thus, cultural theorists, anthropologists, multispecies ethnographers, ecocritics, and others are revealing why "persons" who move in the cosmic realm are associated with a hope for the future, which is being termed "cosmovisions, thousands of years in the making" (Universal Declaration of the Rights of Mother Earth [UDRME] 2010, "Preamble"; see also Introduction, this volume). This hope and vision is playing an important role in the development of a twenty-first century movement being referred to as "cosmopolitics," which is taking form around the world in multiple localized movements; for example, the Zapatista movement (see Chapter 11, this volume), Idle No More (see Chapter 5, this volume), and the movement to stop development in the Alaskan National Wildlife Refuge (see Chapter 3, this volume). As articulated in the Universal Declaration on the Rights of Mother Earth and Climate Change (UDRME), drawn up at the World People's Conference on Climate Change and the Rights of Mother Earth held in Bolivia in 2010, representatives of Indigenous, Global South, and multi-ethnic groups from around the world proclaimed that "violations against our soils, air, forests, rivers, lakes, biodiversity, and the cosmos are assaults against us" ("Final Conclusions" 2010). "Us" is emphasized to mean all living "beings," not just Indigenous peoples themselves. Delegates to this conference and many other Indigenous-led meetings are advocating a coalitional politics that is taking the form of new legislation or revised constitutions (for example, in Bolivia, Ecuador, Peru, and New Zealand). For example, Ecuador's constitution, revised in 2008 to include Title VII, often referred to as the *Bien Vivir* or "Good Way of Living" chapter, expands the concept of human rights to include rights for the intergenerational, evolutionary space and time required, not just for the survival of humans, but all species, both human and nonhuman (Ecuador Constitution. 2008).

In this chapter, to illustrate the role of transformational beings in movement politics, Joni Adamson and Juan Carlos Galeano collaborate on an analysis and explication of six poems from Galeano's chapbook, *Yakumama and Other Mythical Beings,* and six paintings by Columbian artist Solmi Angarita that illustrate the six poems.[5] They explain how the six beings at the center of Galeano's poems act as "seeing instruments" that offer listeners a "complex navigational system" for moving "beyond the present" (Adamson 2001, 141, 145).[6] Galeano's words and Angerita's art offer readers imaginative entry into Amazonian philosophies of "multinatural perspectivism," and at the same time illustrate why these beings are becoming more and more important in ecocriticism, material ecocriticism, multispecies ethnography and the cosmopolitical movement. The chapter concludes with a reflection on Galeano's work as a "participant observer" helping to pilot waste and water management and food security projects in the small town of Padre Cocha, on the Nanay River in the Peruvian Amazon. It offers one example of small-scale cosmopolitics that is bringing together a coalition of artists, academics, and community members to plan and implement solutions to local social and environmental challenges. The importance of Galeano's project (like other projects being piloted by international networking initiatives, such as "Humanities for the Environment") will also be discussed below. Each of these projects is illustrating how stories, ancient and modern, can be employed as "seeing instruments" for building desirable and livable futures.

Yakumama

The Amazon River is often referred to as Yakumama. Like other oral traditions emerging from Indigenous cultures around the world, Amazonian oral traditions represent significant geographic features or ecological systems, to use the words of Julie Cruickshank, that function as an "intensely social space where human behavior, especially casual hubris or arrogance, can trigger dramatic and unpleasant consequences in the physical world" (2005, 11). The people interviewed in Richard and Galeano's film, *The Trees Have a Mother* (2008), clearly see the river as a multinatural space upon which their lives and well-being depend. In "Yakumama," Galeano represents the river as a "serpent" or "Mother" that keeps the "fish happy." Angarita paints Yakumama as a great green anaconda, or *Cobras Grandes*, alive with the activities of both humans and nonhumans (Figure 12.1).

Figure 12.1 "Yakumama." Source: Courtesy of Solmi Angarita.

Yakumama, by Juan Carlos Galeano

The only serpent that creates rivers to keep the fish happy.
Anacondas, Mãe d'Água, Cobras Grandes, and Puraguas are in the same family tree.
Every day she takes the fish out to see the sights, takes them out to eat in out-of-the-way neighborhoods.
Sometimes she travels as a ship lit up at night, people dancing, raising their glasses.
Or she disguises herself as a famous actress in flashy clothes.
When the Yakumama so desires, clouds form along her pathway.
It's hard to imagine the life of some cetico trees without the Yakumama's tenderness.
So delicate, if the Yakumama is not well treated she'll simply take her leave.

The poem presents Yakumama as a being, who, like a snake, can slip into and out of desirable shapes, such as a "famous actress in flashy clothes" who then takes her fish "to eat in out-of-the-way neighborhoods." The humans living and traveling along her serpent back may be happily dancing and singing in boats, or conversely, in their hubris, overfishing and polluting her waters. They might engage in behaviors thousands of miles away, such as the damming of her waters for electricity, and these activities might result in a warming global atmospheric or oceanic conditions that chase away the "clouds that once formed along her pathway." Yakumama is delicate and, like the clouds, she may "simply take her leave."

As has been well documented by scientists and the media, the Amazon River is indeed disappearing. Local residents interviewed in *The Trees Have a Mother* (2008) fear the river is being turned into "a ditch." They express concerns about human activities, including the effects of noisy ships and outboard motors on the animals in the river, the consequences of extensive logging, and effects of garbage and pollution flowing into the river. Indeed, the Amazon River is already "taking her leave." Drought, caused by warming oceanic and atmospheric temperatures linked to climate change, is predicted to reduce the levels of the rivers in the watershed by up to 85 percent in the next ten years. Freshwater dolphins, manatees, and fish that depend on this riverine ecology are already in serious decline (Adams 2009). Tales about Yakumama, then, are still relevant today, working as "seeing instruments" that allow humans to make comparisons between past, present, and future conditions of a river on which all their lives, and those of many nonhuman others, depend.

Mapinguari

Other "persons," such as Mapinguari, might be a friend to monkeys and other creatures of the forest, but enemy to those who exploit Sachamama, Mother of the Forest. Angarita paints this being as an angry, one-eyed ogre. (Figure 12.2) Galeano's poem about Mapinguari acts as an implicit critique of the colonial legacies, economic narratives, and consumption patterns that are contributing to the retreat of Sachamama.

Figure 12.2 "Mapinguari." Source: Courtesy of Solmi Angarita.

Mapinguari, by Juan Carlos Galeano

So the animals might find a day of rest, the Mapinguari uses his single eye to go after those who hunt on Sundays.
Even garlic and tobacco gifts can't placate this ogre's rage.
As for those who hunt for whatever reason—their rifles, arms, and compasses disappear through his mouth which spills straight into his stomach.
Not even an instant to think about his family; not a second to tell the dog goodbye.
The Mapinguari only smiles if tourists give him marshmallows.
After his Sunday patrol, the Mapinguari has the remainder of the week to take his rest.

This poem suggests that Sachamama deserves, metaphorically, a Sabbath or day of rest. Humans may venture into the Amazon rainforests on a Sunday only if they desire to feed marshmallows to forest creatures. Others with more exploitative desires will face Mapinguari's wrath. The ogre will ingest the hunter's guns, the fisherman's boats, and the tourist's compass. Mapinguari is not opposed to humans making their economic living and supporting themselves in the forest most days of the week; however, if Yakumama is to survive, humans must check their greed and see their survival as dependent on, and completely entangled with, the life and well-being of the rainforest. Richard and Galeano's film, *The Trees Have a Mother* (2008), illustrates how Amazonians use stories about beings such as Mapinguari to address their own anger that the big fish they once depended on for food "are gone, the water is poisonous, and the children are dying from contamination."

This is the space from which the cosmopolitical movement emerges. Mapinguari suggests, as does the Universal Declaration on the Rights of Mother Earth and Climate Change (2010), that humans need to become protectors of the forests and rivers. They must, as suggested by the revised Constitution of Ecuador (2008), live in a "good way" and expand the concept of human rights to include rights for the intergenerational, evolutionary space and time required not just for the survival of all species on Mother Earth.

Curupira

One of the "most talked-about protectors of nature" in *Amazonia* is a more-than-human plant-being that blurs species and gender lines. Curupira sometimes appears in the form of a tree or in the form of a girl, described as having long blonde hair that looks like the vines that commonly hang from the trunks of the great trees in the rain forest (Galeano 2009, 41, 42). Angarita's painting shows Curupira with "one foot pointing ahead and the other pointing behind." (Figure 12.3) Galeano's poem suggests that feet pointing in opposite directions gives this being a multiplicity of roles, generous giver of bounty for those respectful of the forest's bounty, or wise protector from those who come into the forest without understanding the secrets that must be understood if the whole forest is to survive.

Figure 12.3 "Curupira." Source: Courtesy of Solmi Angarita.

Curupira, Juan Carlos Galeano

With one foot pointing ahead and the other pointing behind, the Curupira walks through the jungle and tends to the animals, braiding the young palm trees' hair.
Hunters offer the Curupira cigars in exchange for his secrets.
The Curupira puffs the cigars; animals, trees, and fruits appear in the path of his smoke.
But the men shouldn't make off with all the animals, trees, and fruits.
The Curupira could blow smoke so the animals, trees, and fruits disappear.
Blow all his smoke and make the paths vanish.
He could also tell the animals his secrets for hunting men.

In *The Trees Have a Mother* (2008), local residents employ Curupira to theorize about the complex forces of globalization. Those interviewed suggest that Curupira, the being who once protected animals, trees, and fruits from overzealous harvest, is being driven away by loggers, miners, oil companies, agribusiness enterprises, and cocaine producers who do not "ask permission from the Mothers of the Trees" before they destroy "the whole forest." Like Mapinguari, Curupira illustrates, to use Cruikshank's words, that cosmological stories are still relevant today as they act to provide a "set of principles for thinking about current events" (Cruikshank 2005, 57).

Yaras

Playful and social, pink freshwater dolphins were once numerous in the Amazon River and its tributaries but are disappearing because of increased fishing and tourist activities. They figure commonly in oral traditions. For example, in one story, a woman named María disappears for a time then returns to give birth to twin baby dolphins. Her relatives simply take it in stride (Galeano 2009, 19–20). Yakumama, the giant anaconda, presides over "cities" of these dolphins, described alternately as "gringos," "policemen," or "rich people" living at the bottom of rivers. Similarly, "yaras," or manatees, have long been seen sunbathing leisurely along river shores, but their numbers are also dropping. They are thought of as "persons" who live normal, if somewhat luxurious lives, under the water, and sometimes show up in stories as lovers or seducers. In Angarita's illustration, a manatee is seductively represented as "half woman, half fish." (Figure 12.4) However, those who are greedy should beware, since a seductive creature, writes Galeano, might be a purposely menacing presence to those who care only for "riches in the jungle" (Galeano 2009, 49).

Figure 12.4 "Yaras." Source: Courtesy of Solmi Angarita.

Yaras, by Juan Carlos Galeano

Half woman, half fish, Yaras seduce men in search of adventure and riches in the jungle.
Yaras live to sing and don't have to cook.
(Biologists argue that they are merely manatees sunbathing along the shores.)
Men inflict all kinds of wounds on the rivers—wounds cured by songs of the Yaras.
The Yaras offer easy love at any moment to fishermen in need.
Those who taste the Yara's honey stay in the jungle forever.
Along any river, a woman singing or combing her pubis could be a Yara.

"Men inflict all kinds of wounds on the rivers" and their activities, including luxury tourism and overfishing that feeds a world market, are causing the decline of dolphins and manatees. However, the poem suggests hope; these wounds can be "cured by songs." This singing, or the common cacophonous sounds of Amazonian animals, calls upon humans to shift their attention, in the language of the Working Group on Indegenous Peoples at the World Conference on Climate Change and the Rights of Mother Earth, from the greed of living "better" to the idea of "living well." The Final Conclusions of the Working Group specifically delineates a difference between monetary or material riches, from "living well," which is defined as human behaviors that support "a society based on social and environmental justice, which sees life as its purpose" ("Final Conclusions" 2010). The songs of yaras raise questions about what it would mean to make "life" the purpose of society and health and well-being of all creatures the goal of all future human activity.

Renacos

Renacos are Amazonian ficus or banyon trees. This species beautifully illustrates the notion of a "person" as a "collected thing." In "Renacos" (Figure 12.5), Angarita paints the roots of these trees, which tangle thickly with the roots of other Renacos, growing along the banks of the Amazon, bare and gnarled, providing habitat for plants, animals and fish and protection from erosion along the banks of the river. In tales of these trees, and in Galeano's poem, Renacos are "companions" to other Renacos as well as "spirits who protect" all the creatures who "rest or lay eggs" in the tangled roots growing along the banks of rivers (Galeano 2009, 94). These spirits are said to cause fear since they are known to become angry towards any humans who might damage the critical habitats of sheltering young fish or birds.

Figure 12.5 "Renacos." Source: Courtesy of Solmi Angarita.

Renacos, Juan Carlos Galeano

For love or simply to find company, the Renaco tree searches for another tree and grows beautiful thanks to the life of its new companion.
A tree able to fall in love with another must have special qualities.
Without caring what people say, Renacos live just to embrace other trees.
"Tree killers, ingrates," they are called by those who don't know them and accuse them of sucking the blood of everyone else.
Because of their bare roots, people call them tree bandits.
But those with crushes and desperate lovers wish the Renacos' juices could glue them together.
Those who say bad things about Renacos are merely jealous.
Anacondas and boas who sing lullabies in their branches understand the Renacos and find truth in their philosophy.

Renaco trees are wonderful examples of beings who represent "collected things" or, in other words, the entangled, multi-generational systems at work in forests, mountains, or rivers. When their material existence and that of the worlds to which they belong is threatened by humans who "don't know them" or don't understand "their philosophy," Renacos as "protector spirits" might be marshaled, along with the humans and nonhuman who may live in or beside them along river banks, into a movement to protect a river, or the entangled plants protecting the water and fish from erosion. A multiplicity of life is held in their "embrace." Trees may be categorized as "natural resources" within Western politics or science, but in Ecuador, Bolivia, and Peru, for example, Indigenous activists and politicians are confronting the monopoly of discourses that provincialize "the universe" as a world inhabited by humans who are distanced from "Nature" (de la Cadena 2010, 345). Cosmopolitical actors, and the cosmological beings that support them, resist the exploitation of multigenerational social and ecological systems. These actors are organizing around the notion—as articulated in the UDRME and other legal instruments such as the revised Constitution of Ecuador—that multiple species and "things" that have heretofore not been considered deserving of the same rights and protections as humans should now be protected by humans who understand their "philosophies" and by legislation granting them rights to continue to exist and thrive (see Haraway 2015, Vivieros de Castro 1998; Vivieros de Castro 2004).

Matinta Perera

Stories stemming from the Amazonian imagination have long inspired art, filmmaking, music, literary expressions, and other means of representation, but they also offer listeners feelings of optimism and hope for survival in the midst of "the crisis of the global world" (Leff 2010).

Galeano's poems and Angarita's paintings convey the element of humor commonly associated with tales about transformational beings—from

Coyote in the North to Curupiri in the South. Often, the humor in these tales offers insight into the ordinary occurrences taking place in every community daily. For example, in "Matinta Perera," Galeano illustrates how a "person" might seem as ordinary as a "devious" neighbor who might be male, or perhaps female, or perhaps, at night, a bird (Galeano 2009, 73–74). As Angarita imagines "Matinta Perera," this neighbor might take the form of a brilliantly azure feathered human/animal making too much noise and causing others to lose sleep (Figure 12.6).

Figure 12.6 "Matinta Perera." Source: Courtesy of Solmi Angarita.

Matinta Perera, by Juan Carlos Galeano

A town-dweller, who wears the clothes of a bird for a little night flying.
His shrill songs keep people from sleeping.
(When he appears on the streets dressed like a modest old lady, the Matinta is harmless.)
Because some terrible thing or other has happened or some light goes missing, the Matinta's whistle pierces the walls and the hearts along the riverbanks.
If only his beak were painted with lipstick and he was wearing high heels, he'd be ready for the party.
But he screeches as if he had thorns for feathers.
The Matinta Perera only calms down if someone steps outside and shouts: "Compadre, come back early tomorrow morning. We'll have a cup of coffee and a little tobacco for you."
The Matinta would rather croon but his voice is more hindrance than help.

As Galeano explains, stories about Matinta Perera, a common bird, are often employed as a thought experiment about people, either harmless, or perhaps, more harmful, living closely together (2009, 74). Most humans, living in close proximity to others, must deal at some point in life with a noisy neighbor, whether that person takes the form of a "modest old lady" or a more devious person out for a little "night flying." In such a situation, one might employ humor, imagining the modest neighbor as a flamboyant woman, in red high heels. Or the "Matinta Perera" might help one think of ways to deal with a more malicious person, one that might have "thorns for feathers." Humor, in ordinary or dangerous situations, offer feelings of relief, optimism, and hope for resilience even in the face of potential crisis.

Small Scale Cosmopolitics, Humanities for the Environment

In phone conversations with Adamson, Galeano has observed that he remembers that when he was a child, government trucks would pull up to rivers to dump loads of garbage. Today, garbage is an even larger problem, still being dumped into the rivers, and then moving through communities, polluting the forests, and causing illness to humans, fish, animals, and plants before flowing into the great serpent, Yakumama, the Mother of All Water Beings (Fernandez 2015, n.p.). This urgent problem has become Galeano's motivation to create a clean-up program in the small town of Padre Cocha on the Nanay River, in the Peruvian Amazon, where he filmed *The Trees have a Mother*. Drawing on his reputation as a poet, he is organizing environmental engineers, sociologists, artists, writers, and his students into a collaboration that will work to solve the waste management and food security problems in Padre Cocha. Like the delegates who gathered in the Bolivia for the World People's Conference on the Rights of Mother Earth, the local people of Padre Cocha, together with Galeano and his students, are aiming to keep garbage from entering the region's rivers. Their goal is to build "a society based on social and environmental justice, which sees life as its purpose" ("Final Conclusions 2010, Common Objective 2). The project is now in the kickstarter phase, but Galeano and his collaborators are planning their project as a template that might be implemented in other Amazonian communities that might replicate and scale the plan for their own locations (Fernandez 2015, n.p.).

Similarly, Adamson has been involved in an international project funded by the Andrew W. Mellon foundation called "Humanities for the Environment" that is creating template projects that are available open access for other communities to replicate and scale.[7] For example, the *Archive of Hope and Cautionary Tales* is a digital collection of stories about frontline communities organizing to advocate, like Padre Cocha, for "a society based on social and environmental justice, which sees life as its purpose" (Humanities for the Environment 2013). Like Galeano, the researchers and artists contributing to this project are implicitly raising the question: "*How* do the humanities and the arts contribute uniquely through storytelling to solving social justice and environmental challenges?" Galeano's poems and Angarita's paintings suggest answers to this question. Similarly, the stories in the Archive of Hope and Cautionary Tales seek to test and rethink how communities might employ imagination and story, and digital tools, to make the knowledges of local Indigenous and non-Indigenous communities more widely available. For example, in "Renewing Relatives: Nmé Stewardship in a Shared Watershed," philosopher Kyle Powys Whyte (Citizen Potawatomie Nation), Mary Holtgen (Little River Band of Ottawa), and Stephanie Ogren (Little River Band of Ottawa), make Indigenous stories about the Nmé, or sturgeon, central to the Little River Band of Ottawa's program to restore a threatened fish to the Big Manistee River Watershed in Michigan (Holtgen 2015). This fish is not just an iconic species in the Great Lakes region, it is a species sacred to the Ottawa because it so central to the collected or entangled lifeways of an Indigenous group and one of the species upon which it has depended for centuries. "Renewing Relatives" tells the story not only of how an Indigenous community is sharing a watershed with its relative, the sturgeon, but collaborating with the surrounding non-Indigenous communities members who live there as well. In this way, they are teaching an entire community about the "philosophy" of the sturgeon, just as Galeano and Angarita are sharing the philosophies of Yakumama, Yaras, and Renacos.

The stories in the Archive can also be read as "cautionary tales" since they illustrate that no one solution should be proposed as a panacea. As Mike Hulme, a climate scientist who contributed research to the United Nations' Intergovernmental Panel on Climate Change Reports has observed, framing complex environmental changes as "mega-problems" necessarily demands "mega-solutions" and this "has led us down the wrong road" (Hulme 2009, 332). In contrast to research that proposes "mega-solutions," Galeano's Padre Cocha project and the Little River Band of Ottawa's conservation plan for the sturgeon illustrate how humanists and artists are working with small community-based alliances to seek not one plan of action, but a range of evidence-based, reasoned, scaled and culturally diverse responses "reflective of life in a plural world" (Castree et al. 2014, 765–66).[8] They employ stories about transformational beings and collected things as thought experiments, or tools, that might be employed to build livable, sustainable futures.

Notes

1. Momaday and Harjo were Adamson's professors. See also the "Politics and Poetics" seminar where she studied with Simon Ortiz and Leslie Marmon Silko in 1992 at the University of Arizona, http://poeticsandpolitics.arizona.edu/series/poetics-and-politics-1992.
2. For more on the "participant observer" methodologies developed by Franz Boas and his students (including Zora Neale Hurston) and its deep roots stretching back to Alexander Von Humboldt's travels to the Americas in 1799 ce, see Adamson 2013, Adamson 2016, and the Introduction to this volume by Adamson and Monani.
3. For more on Adamson's work with Native North American students, see Adamson (2001, xiii–xx).
4. For more information on the conference, see https://osmilnomesdegaia.eco.br.

5. With a first edition published in 2014, Galeano is still expanding *Yakumama and Other Mythical Beings* with additional poems and will publish a new edition, featuring these illustrations by Angarita, in the near future.
6. Adamson's formulation (2001) of "seeing instrument" is based on anthropologist Dennis Tedlock's explication of the work of transformational gods found in the Popol Vuh. These beings and the stories about them, and the Popol Vuh itself, are thought of as "seeing instruments" for understanding the present and making decisions for the future (Tedlock 1985, 32, 71).
7. Adamson is lead web developer of the international Humanities for the Environment (HfE) Andrew W. Mellon project, see http://hfe-observatories.org; she is also a Principle Investigator for the North American Observatory of the Humanities for the Environment project, see http://hfe-observatories.org/observatories/north-american-observatory/.
8. See Holm, et al., "Humanities for the Environment – a Manifesto for Research and Action," (2015) written to guide a second phase of the Humanities for the Environment project and to affirm the expertise of the humanities and of storytelling, for transforming human behaviors, preferences, practices, and actions in a time when there is a need for radical change.

References

Adams, David. 2009. "Amazon Could Shrink by 85% Due to Climate Change, Scientists Say," *The Guardian*. March, 11. Accessed January 2016. http://www.guardian.co.uk/environment/2009/mar/11/amazon-global-warming-trees.

Adamson Clarke, Joni. 1992. "Why Bears are Good to Think and Theory Doesn't have to be Another Form of Murder: Transformation and Oral Tradition in Louise Erdrich's *Tracks*." *Studies in American Indian Literatures* 4 (1): 28–48.

Adamson, Joni. 2001. American Indian 2001. *American Indian Literature, Environmental Justice, and Ecocriticism: The Middle Place*. Tucson: University of Arizona Press.

———. 2013. "Environmental Justice, Cosmopolitics and Climate Change." *The Cambridge Companion to Literature and Environment*, ed. Louise Westling, 169–83. New York: Cambridge University Press.

———. 2016. "Humanities." In *Keywords for Environmental Studies*, ed. Joni Adamson, William A. Gleason, and David N. Pellow, 135–39. New York: New York University Press.

Babcock, Barbara A. 1975. "Why Frogs Are Good to Think and Dirt Is Good to Reflect On." *Soundings* 58 (2): 167–81.

Castree, Noel, William M. Adams, John Barry, Daniel Brockington, Bram Buscher, Esteve Corbera, David Demeritt, Rosaleen Duffy et al. 2014. "Changing the Intellectual Climate." *Nature Climate Change* 4 (9): 763–68.

Cruikshank, Julie. 2005. *Do Glaciers Listen? Local Knowledge, Colonial Encounters, and Social Imagination*. Seattle: University of Washington Press.

de la Cadena, Marisol. 2010. "Indigenous Cosmopolitics in the Andes: Conceptual Reflections beyond 'Politics.'" *Cultural Anthropology* 25 (2): 334–70.

Ecuador Constitution. 2008. Title VII, Chapter 2, Secs. 1–7, 134–39. https://www.constituteproject.org/constitution/Ecuador_2008.pdf.

Fernandez, Robert. 2015. "Juan Carlos Galeano in Conversation." *Harriet: A Poetry Blog*. Poetry Foundation. 17 September. Accessed January 16, 2016. http://www.poetryfoundation.org/harriet/2015/09/juan-carlos-galeano-in-conversation/.

"Final Conclusions Working Group 7: Indigenous Peoples." 2010. Working Paper. In *World Peoples Conference on Climate Change and the Rights of Mother Earth* [Indigenous Peoples' Declaration]. April 30. Cochabamba, Bolivia: UDRME.

Galeano, Juan Carlos. 2009. *Folktales of the Amazon*. Trans. Rebecca Morgan and Kenneth Watson. Westport, CT: Libraries Unlimited.

———. 2012. *Amazonia*, 2nd (Spanish/English) ed. Trans. James Kimbrell and Rebecca Morgan. Inquitos, Peru: Centro de Estudios Teológicos de la Amazonía (CETA).

———. 2014. *Yakumama and Other Mythical Beings*. Trans. James Kimbrell and Rebecca Morgan. Iquitos, Peru: Tierra Nueva Editores.

Haraway, Donna. 2015. "Anthropocene, Capitalocene, Plantationocene, Chthulucene: Making Kin." *Environmental Humanities* 6 (August): 159–65.

Holm, Poul, Joni Adamson, Hsinya Huang, et al. 2015. "Humanities for the Environment—A Manifesto for Research and Action." *Humanities*. 4 (4): 977–92. doi:10.3390/h4040977.

Holtgen, Mary and Stephanie Ogren, with Kyle Powys Whyte. 2015. "Renewing Relatives: Nmé Stewardship in a Shared Watershed." http://hfe-observatories.org/project/renewing-relatives-nme-stewardship-in-a-shared-watershed/.

Hulme, Mike. 2009. *Why We Disagree about Climate Change: Understanding Controversy, Inaction and Opportunity*. Cambridge, UK: Cambridge University Press.

Humanities for the Environment (HfE), West Cluster. 2013. *Archive of Hope and Cautionary Tales*. November. Tempe, AZ: Arizona State University. http://hfe-observatories.org/projects/hope-and-caution/.

Leff, Enrique. (2010). "El devanecimiento del sujeto y la reinvención de las identidades colectivas en la era de la complejidad ambiental." *Polis, Revista Latinoamericana,* 9 (27): 151–98.

Lévi-Strauss, Claude. 1982. *The Way of the Masks*. Trans. Sylvia Modelsik, Seattle: University of Washington Press.

Tedlock, Dennis. trans. (1985). *Popol Vuh: The Definitive Edition of the Mayan Book of the Dawn of Life and the Glories of Gods and Kings*. New York: Touchstone.

Trees Have a Mother, The: Amazonian Cosmologies, Folktales, and Mystery. 2008. Directed by Valliere Richard Auzenne and Juan Carlos Galeano, Tallahassee: Florida State University Film School. Available at Films on Demand, item number 40566. http://digital.films.com/play/WNHAND.

Turner, Victor. 1967 "Betwixt and Between: The Liminal Period in *Rites de Passage*." *The Forest of Symbols: Aspects of Ndembu Ritual*. Ithaca, NY: Cornell University Press, 93–111.

Universal Declaration on the Rights of Mother Earth and Climate Change (UDRME). 2010. World People's Conference on Climate Change and the Rights of Mother Earth. 22 April. https://pwccc.wordpress.com/programa/.

Uzendoski, Michael. 2009. Foreword to *Folktales of the Amazon*, by Juan Carlos Galeano, trans. Rebecca Morgan and Kenneth Watson, ix–xii. Westport, CT: Libraries Unlimited.

Vidal, John. 2011. "Bolivia Enshrines Natural World's Rights with Equal Status for Mother Earth." *guardian.co.uk*. 10 April. http://www.guardian.co.uk/environment/2011/apr/10/bolivia-enshrines-natural-worlds-rights/print.

Viveiros de Castro, Eduardo Batalha. 1998. "Cosmological Deixis and Amerindian Perspectivism." *Journal of the Royal Anthropological Institute* 4 (3): 469–88.

———. 2004. "Exchanging Perspectives: The Transformation of Objects into Subjects in Amerindian Ontologies." *Common Knowledge*. 10 (3): 463–84.

List of Contributors

Joni Adamson is Professor of Environmental Humanities in the Department of English and Senior Sustainability Scholar at the Julie Ann Wrigley Global Institute of Sustainability at Arizona State University. She has authored and co-edited many books and volumes that have helped to establish and expand the fields of environmental humanities and environmental studies, including *Keywords for Environmental Studies* (2015), *American Studies, Ecocriticism, and Citizenship* (2013); *The Environmental Justice Reader* (2002), *and American Indian Literature, Environmental Justice, and Ecocriticism* (2001). Adamson is author of over fifty articles and reviews focusing on Indigenous perspectives on environmental justice, material ecocriticism, environmental literature and film, food sovereignty, and critical plant studies. She is a Principle Investigator for Humanities for the Environment (H*f*E) and leads the 2.0 development team for the H*f*E international website, http://hfe-observatories.org.

Solmi Angarita studied art at the Universidad Nacional de Colombia, in Bogotá, Columbia. Angarita designs and illustrates books for children and young adults. Since 2011 she has been involved in projects dedicated to education and creativity for youth and children in the Amazon regions of Peru and Colombia. Currently she teaches art technique, drawing, painting, and charcoal art in the Indigenous school, La Garza Blanca, in the riverine town of Pevas, on the banks of the Ampiyacu River in the Peruvian Amazon. Her work, which draws from perspectivism and Amazonian cosmovisions, has been exhibited in galleries such as those at the Universidad Nacional de Colombia, Espacio 101 (Bogota, Columbia), and at the Universidad de Tunja, Columbia.

Rayson K. Alex teaches ecocriticism, cinematic arts, and aesthetics at the Birla Institute of Technology and Science Pilani—K. K. Birla Goa Campus in India. His scholarly interests include eco-indigeneity and ecocinema. In addition to numerous articles, his has co-edited *Essays in Ecocriticism* (2007) and *Culture and Media: Ecocritical Explorations* (2014). Alex is the founder and co-director of the *tiNai* Ecofilm Festival (www.teff.in) and is a member of the Technical Advisory Committee of the National

Council of Science and Technology Communication for the Government of India's Ministry of Science and Technology.

Subhankar Banerjee is an artist, writer, and environmental humanities scholar. He holds the Land Arts of the American West Endowed Chair and is Professor of Art and Ecology in the Department of Art and Art History at the University of New Mexico. In addition to contributing essays in several anthologies and journals, he is author of *Arctic National Wildlife Refuge: Seasons of Life and Land* (2003) and editor of *Arctic Voices: Resistance at the Tipping Point*. His photographs have been exhibited in more than fifty museums and galleries around the world. Professor Banerjee received a Greenleaf Artist Award from the United Nations Environment Programme (UNEP) and a Cultural Freedom Award from the Lannan Foundation of Santa Fe, New Mexico.

Clint Carroll is an enrolled citizen of the Cherokee Nation and is Assistant Professor of Ethnic Studies at the University of Colorado–Boulder. His primary work with Cherokee Nation government and communities in northeastern Oklahoma addresses issues of tribal environmental governance, health sovereignty, and the revitalization of traditional ecological knowledge and practices. During his time at the University of Minnesota (2011–2015), he supported and closely followed tribal opposition to the wolf hunt, learning from local Ojibwe activists and community members, collaborating with nonprofit organizations, and attending gatherings such as the White Earth Wolf Conference in 2013 and the Minnesota Wolf Day Rally in 2014.

Kathryn Yalan Chang is Associate Professor and Chairperson of the Department of Foreign Languages and Literature at Huafan University in Taiwan. Her current interests include nature therapy, material ecocriticism, toxic discourses, environmental justice and activism, and animal studies. Chang's doctorial dissertation is entitled *Nature, Gender, and Risk: Margaret Atwood, Linda Hogan, and Karen Tei Yamashita*. Her latest publications include "An Affective Turn: The Rhetorical Triangle in *The Cove*," in *Culture and Media: Explorations in Ecocriticism* (2014) and "Toxicity Survival: Toxicity, Violence, and Trans-Corporeal Ethics in Indra Sinha's *Animal People*," in *Chung-Wai Literary Monthly* (2015).

Janet Fiskio is Associate Professor of Environmental Studies and Comparative American Studies at Oberlin College in Ohio. Her teaching and research on environmental and food justice are grounded in collaborative and community-based research and teaching in the Rust Belt. In addition, she is active in researching and writing about climate justice and protest through an interdisciplinary methodology. Professor Fiskio has published essays in *American Literature*, *The Cambridge Companion to Literature and the Environment*, and *Race, Gender, and Class*. She serves as the book review editor for *Resilience: A Journal of the Environmental Humanities*.

Juan Carlos Galeano is a poet, translator, and essayist, born in the Amazon region of Colombia. He has published several books of poetry, has translated North American poets into Spanish, and has made a documentary film. His poetry, inspired by Amazonian cosmologies and the modern world, has appeared in college textbooks and has been anthologized and published in international journals such *Casa de las Américas* (Cuba), *The Atlantic Monthly*, *Ploughshares*, *BOMB Magazine*, *Drunken Boat,* and *Poesia*. He lives in Tallahassee, Florida, where he teaches Latin American poetry and Amazonian Cultures at Florida State University.

Pietari Kääpä is Lecturer in Media and Communications and Programme Director of the M.Sc. (Master of Science) in Media Management Program at the University of Stirling in Scotland, UK. His work is focused on both transnational cinema and environmental media. He has published seven books, several articles, and edited four journal issues on these topics. His latest books are *Nordic Genre Film: Small Nation Film Culture in the Global Marketplace* (with Tommy Gustafsson, 2015) and *Ecology and Contemporary Nordic Cinema* (2014).

Angelica Lawson (Northern Arapaho) is an Assistant Professor of Film Studies and Ethnic Studies at the University of Colorado–Boulder. She earned her Ph.D. in American Indian Studies at the University of Arizona, and her book manuscript, *Indigenous Strategies of Resistance and Resilience in American Indian Literature, Film, and Media,* outlines a methodology for utilizing site-specific Indigenous worldviews as a means for determining narratives of resistance and resilience in the work of various Indigenous artists. She has worked with Indigenous peoples from New Zealand, Norway, and Finland and seeks to examine the intersections of the national and transnational in Indigenous film and media.

Salma Monani is Associate Professor at Gettysburg College's Environmental Studies Department. As a humanities scholar her research and teaching include explorations in cine-ecocriticism. She is co-editor, with Steve Rust and Sean Cubitt, of *Ecocinema Theory and Practice* (2012) and *Ecomedia: Key Concepts* (2015). She has also published studies of Indigenous eco-activism, film and environmental justice, as well as film festival studies in journals such as *ISLE (Interdsiciplinary Studies in Literature and Environment)*, *Environmental Communication: The Journal of Nature and Culture*, *Local Environment: The International Journal of Justice and Sustainability* (online), and in various anthologies. She is currently working on a monograph on Indigenous ecocinema.

Shelley Niro was born in Niagara Falls, New York. She is a member of the Six Nations Reserve, Bay of Quinte Mohawk, Turtle Clan. Niro currently lives in Brantford, Ontario. She is a multimedia artist. Her work involves photography, painting, beadwork, and film. Niro is conscious of the impact that (neo)colonial media have had on Indigenous people. Like many artists from different Native communities, she works relentlessly to present people in realistic and explorative portrayals. She graduated from the Ontario College of Art (Honors) and received a Master's of Fine Art degree from the University of Western Ontario. She is the inaugural recipient of the Aboriginal Arts Award presented through the Ontario Arts Council.

Simon J. Ortiz is an acclaimed poet and writer of Acoma Pueblo (North American) heritage. With literary perspectives as a guide, his research and writing interests include the cultural, social, and political dynamics of Indigenous peoples of the Americas. Ortiz's publications in poetry, fiction, creative nonfiction, children's literature and scholarship reflect his varied and active involvement in contemporary Indigenous life and place his name in the canon of world literatures. He is the recipient of many awards, including the White House Salute to Poetry Award, the Pushcart Prize, and the Lifetime Achievement Award from the Native Writers' Circle of the Americas.

Abigail Pérez Aguilera is a Ph.D. in Justice Studies and Social Inquiry earned at Arizona State University. Her dissertation is titled *Nature, Gender and Indigeneity: Three Case Studies of Environmental Movements in Mexico*. She has conducted field research in Mexico City and Oaxaca, Mexico. She writes and researches about contemporary Indigenous movements, literature written by women of color and its connections to environmental social movements, forced displacement, gender violence, and global politics. Her most recent work is in *Environmental Justice and Cultural Representation in Latin America* (forthcoming 2016).

Janine Randerson is an Aotearea/New Zealand-based media artist, writer, and curator. She is currently Program Leader, Master of Performance and Media Arts, at Auckland University of Technology in New Zealand. A research thread in Randerson's work is technological mediation in ecological systems, with a particular focus on the atmospheric sciences. Her current projects situate media art in relation to water, weather, and politics, both locally (*Other Waters: Art on the Manukau, Auckland* [2014]), as well as internationally (*Heat Islands,* for the 2015 Prague Quadrennial: Shared Space: Weather, Music and Politics). The website for Janine Randerson's media art practice is http//: www.janineranderson.com.

Laura Tohe is Diné. She is Sleepy Rock Clan, born for the Bitter Water clan. A librettist and poet, she is the author of *Making Friends With Water* (1986), *No Parole Today* (1999), *Tseyi, Deep in the Rock: Reflections on Canyon de Chelly* (2005), and *Code Talker Stories* (2012), an oral history book on the Navajo code talkers. She is also editor, with Heid Erdrich, of *Sister Nations: Native American Women Writers on Community* (2008). Her libretto, *Enemy Slayer, A Navajo Oratorio*, had its world premiere in 2008 with the Phoenix Symphony Orchestra, which was recorded on the Naxos classical music label. She is Professor with Distinction in Indigenous Literature in the English Department at Arizona State University. She is Navajo Nation Poet Laureate (2015–2017).

Amanda Yates explores issues of ecology and Indigeneity through performative architectures and installations. Yates's work has been presented in the book by Stuart Harrison, *46 Square Metres of Land Doesn't Normally Become a House* (2011) and the journal *Dwell*, as well as at international exhibitions such as the Prague Quadrennial (2015) and IDEA16: Boston (2016). She has contributed chapters to *Architecture in the Space of Flows* (2012) and *Taone Tupu Ora: Indigenous Knowledge and Sustainable Urban Design* (2010) and has coedited the December 2011 special issue of the journal *Studies in Material Thinking*. Yates directs the Emergent Ecologies Lab at Auckland University of Technology, a creative research studio for ecology, Indigeneity and urbanism. She is curatorial director of the Oceanic Performance Biennale, an event that brings creative practice and activism together as it spotlights ecological issues particular to the Pacific region. The website for Emergent Ecologies is: http://www.emergentecologies.net.

Index

For Product Safety Concerns and Information please contact our EU
representative GPSR@taylorandfrancis.com
Taylor & Francis Verlag GmbH, Kaufingerstraße 24, 80331 München, Germany

www.ingramcontent.com/pod-product-compliance
Ingram Content Group UK Ltd.
Pitfield, Milton Keynes, MK11 3LW, UK
UKHW022001180926
475701UK00003B/143

* 9 7 8 0 3 6 7 6 6 8 1 2 9 *